LIBRARY
HARPER ADAMS AGRICULTURAL COLLEGE

D1582958

LONG LOAN

RETURNED

WITHDRAWN

INSECT PESTS

of

Small Grains

Wendell L. Morrill
Montana State University,
Bozeman

APS PRESS
The American Phytopathological Society
St. Paul, Minnesota

ADAMS AGRICULTURAL LIBRARY
632.7 MOR
63478
COLLEGE

Reference in this publication to a trademark, proprietary product, or
company name is intended for explicit description only and does not imply
approval or recommendation to the exclusion of others that may be suitable.

Library of Congress Catalog Card Number: 95-76088
International Standard Book Number: 0-89054-200-7

© 1995 by The American Phytopathological Society

Photo credits
Photographs are copyrighted as noted: color plates 1–94 and figures
2.1–2.3, 4.1–4.18, 5.5–5.8, 15.1, and 21.1–21.4, copyright W. L. Morrill;
figures 9.1–9.7 and 10.1–10.6, copyright Montana State University.

All rights reserved.
No portion of this book may be reproduced in any form, including photo-
copy, microfilm, information storage and retrieval system, computer data-
base, or software, or by any means, including electronic or mechanical,
without written permission from the publisher.

Printed in the United States of America on acid-free paper

The American Phytopathological Society
3340 Pilot Knob Road
St. Paul, Minnesota 55121-2097, USA

Preface

This book was written for hail adjusters, growers, county agents, consultants, and students. It will be useful for diagnosis of crop symptoms and identification of pest insects commonly associated with wheat, oats, and barley in the United States and Canada. The information herein is a compilation of current literature and personal experience with small grain insect pest management during the past 30 years.

Photographs in this book are copyrighted as noted on the copyright page and may not be reproduced without permission. Additional information concerning trap construction and future releases of software on small grain insects may be obtained from the author.

I wish to acknowledge mentors who have influenced my career. Robert Walstrom and Robert Kieckhefer supervised my undergraduate and graduate training at South Dakota State University and at the Northern Grain Insect Research Laboratory near Brookings. Further research at the graduate level at the University of Florida was supervised by Gerald Greene. Ham Tippins provided advice and encouragement during development of grassland and small grain insect projects at the University of Georgia. B. Merle Shepard offered the opportunity to study pest insects at the International Rice Research Institute in the Philippines.

Early versions of selected chapters were reviewed by Phil Bruckner, Keith Pike, Robert Kieckhefer, Harold Toba, and Jay Karren. Later versions were reviewed by Mike Weiss and Jerry Michels.

Wendell L. Morrill

About the Author

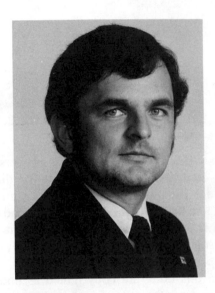

Wendell L. Morrill is a professor of entomology at Montana State University. He has worked with insect management in wheat, corn, rice, alfalfa, and grasslands for more than 30 years. His research has emphasized refinement of insect-monitoring methodology, crop loss assessment, and control strategies and has been conducted across the United States and in the Philippines. His field research has been closely coordinated with growers, consultants, and county agents.

He received undergraduate and master's degrees in entomology from South Dakota State University and a doctorate in entomology from the University of Florida. He grew up on a farm in South Dakota, where he obtained a good background in small grain production and an appreciation of the need for maintaining profit margins.

Contents

INSECT
PESTS
of
Small
Grains

Chapter 1

Early History of Agriculture and Insects

The first foreign insects of agricultural importance that were introduced by man into North America probably were carried from the Old World by fishermen during the 16th century. Their ships commonly carried rock and soil ballast that probably was infested with plants and soil-inhabiting insects.

Early North American settlers from Europe brought domestic plants and seeds and also inadvertently introduced weed and insect pests. Newfoundland was the center for the new settlements and therefore became the point of introduction for many of the exotic insect species. The Canadian maritime provinces were settled during the 17th century and were followed by settlement of inland prairie provinces during the 19th century.

In the United States, farming in the prairie states began during the late 1800s. Settlement and agricultural development were encouraged by the transfer of land from public to private ownership by the Homestead Act in 1862 and under the Enlarged Homestead Law in 1909.

Early producers enjoyed a period of adequate moisture, fertile soil, and an absence of serious weed and insect pests. The land prices were low, grain yields were plentiful, and prices for agricultural products were high. However, the first widespread drought occurred during 1917–1920 and was accompanied by grasshopper outbreaks and Russian thistle infestations. This drought provided a hint of even greater problems that were to follow during the major dust bowl drought years of the 1930s.

The North American wheat belt was developed in the original

prairie grassland region that extended from Indiana southwest into Texas and northward into Alberta, Saskatchewan, and Manitoba. This region was characterized by the eastern tall-grass prairie that gradually gave way to short-grass prairie in the west and fescue grassland in the north. Many native insects, including wireworms, cutworms, and grasshoppers, quickly adapted from native grasses to small grains.

Some native insect species adapted more slowly to cultivated crops. The wheat stem sawfly was originally present in native grasses and appeared in spring wheat after about 10 years. After 70 years, it had adapted to winter wheat as well. During the last 100 years, about six of the more than 100 species of cutworms known to occur in the northern Great Plains have become economically important pests in small grains. How many more will adapt during the next 100 years?

Numbers of some species of native insect pests may have declined as a result of man-made changes in the environment. For example, the Rocky Mountain grasshopper was a major pest of crops in Manitoba during the 19th century but is now believed to be extinct. These grasshoppers laid eggs in dust wallows that were created by buffalo. Disappearance of the buffalo and their wallows reduced the availability of reproduction sites and may have been an important factor in the extinction of this grasshopper species.

Insects selectively fed on wild grasses and forbs. Loss of these plants caused by intensive grassland grazing by domestic livestock undoubtedly influenced the prevalence of groups of native insect species.

Some introduced species of insects also have become economically important pests of small grains. These include the Hessian fly, wheat midge, and cereal leaf beetle.

Changes in agricultural practices during the past century (Table 1.1) have resulted in changes in insect population densities. The practice of summer fallowing was developed to allow use of precipitation for a 2-year period in a single growing season. The unplanted

Table 1.1. Development of sustainable agricultural practices in the northern Great Plains

Sustainable practice	Purpose	Year of Initiation
Alternate-year summer fallow	Store soil moisture	1920
Strip cropping	Reduce soil erosion caused by wind	1922
Flex cropping	More efficient use of soil moisture	1971
Reduced tillage	Increase snow retention and reduce soil erosion from wind and water	1982

or fallow fields do not have plant foliage necessary to support some species of small grain insect pests.

The rigid alternate-year summer fallow regimes are currently being modified to provide a more flexible cropping system. During years of adequate moisture, some fields that normally would be fallow are now planted. This also reduces the problem of saline seep, a situation in which excess subsoil moisture resurfaces, evaporates, and leaves a deposit of salt on the soil. These salty areas appeared during the 1940s, and by 1971, an estimated one-quarter of a million acres of agricultural land were affected.

The practice of strip cropping, dividing fields into narrow areas that are oriented across the prevailing wind direction, originated during the drought years of 1917–1920. The close proximity of narrow strips of stubble to standing crops enhances survival of stubble-infesting insects that overwinter in the stubble and disperse to crops during the growing season. These pests include the wheat stem sawfly, Hessian fly, and wheat jointworm.

Conservation tillage practices have recently been developed to maximize crop residue on the soil surface and minimize erosion from wind and rain. Deep plowing of crop residue is discouraged, and the need for weed control with herbicides has increased. The long-term effects of conservation tillage and sustainable practices on pest and beneficial insects in small grains are yet to be determined.

Mechanization of agriculture has affected the prevalence of insect problems. For example, oats are resistant to many insect pests that attack wheat. Oats were widely grown for horse feed, but when horses were replaced by tractors, the need for oats declined. Oats were no longer included in rotation systems.

The size of small grain-producing farms has increased. The average wheat-producing farm in Montana was about 5,500 ha in 1993. Increased farm size has mandated longer harvest periods and increased the period of time during which crops are vulnerable to wind, rain, and hail.

There have been many dynamic changes in agriculture since vast acreages of native prairie sod were replaced with small grain fields. We have seen improvements in crop cultivars, more application of fertilizers, registration of safer pesticides, computerized irrigation programs, and improved harvesting equipment. Global positioning technology for spot treatment of problems is currently being developed. These and future practices will continue to affect insect pest management in future cropping systems.

Selected References

Bauder, J. W., ed. 1987. A century of action. Proc. Annu. Meet. Soil Conserv. Serv. Soc. Am., 42nd. Artcraft, Bozeman, MT.

Danks, H. V. 1978. Canada and its insect fauna. Mem. Entomol. Soc. Can., no. 108.

Ford, G. L., and Krall, J. L. 1979. The history of summer fallow in Montana. Mont. Agric. Exp. Stn. Bull. 704.

Morrill, W. L. 1983. Early history of cereal grain insect pests in Montana. Bull. Entomol. Soc. Am. 29(4):24-28.

Pratt, L. H., and Lund, C. E. 1990. Montana Agricultural Statistics. Montana Agricultural Statistics Service, Helena.

Chapter 2

Plant Growth Stages

Cereals are grown for their edible starchy seeds and provide a large portion of human subsistence in the world. The major food grains for human consumption are wheat, rice, oats, maize (corn), rye, grain sorghum, and barley. Corn, barley, oats, and sorghum are also produced as feed grains for animal consumption. This book refers to wheat, barley, and oats as small grains.

Small grain plants pass through several stages during development from seeds to mature plants. Insect activity is seasonal, and the plant growth stages vary in resistance to insects and subsequent yield loss. From the entomological viewpoint, the plant growth stages can be categorized into the seed, seedling, tillering, stem elongation (includes jointing and booting), heading, and ripening periods (Table 2.1).

Growth Stages

Seed

The seed stage lasts about 5 days from planting until germination. The period may be extended because of lack of moisture or low soil temperature. Insects known to attack seeds include wireworms, false wireworms, and seed maggots.

Seedling

The seedling stage begins with emergence of the root (radicle) and shoot (coleoptile) from the seed and continues until additional tillers appear. Insects that attack seedling plants include wireworms, cutworms, aphids, grasshoppers, and cereal leaf beetles. The growing point is underground, and plants can continue to grow even though aboveground vegetation is consumed by attacking insects.

Tillering

Thirty or more stems may develop from the same plant when growing conditions are favorable. Fall-planted small grain tillers appear before the cold winter season begins, and tillering continues during spring growth. Spring-planted grain produces tillers for 2–3 weeks. The tillering stage terminates when stem elongation begins.

Insects that infest small grain plants in the tillering stage include wireworms, cutworms, aphids, grasshoppers, Hessian flies, cereal leaf beetles, and wheat stem maggots. Hail and insect damage increases the amount of tillering. Roots are well established when tillers appear, and the growing point is underground. Plant regrowth can occur after defoliation by grasshoppers or cutworms.

Stem Elongation

During this stage, stems elongate and the growing point that produces the nodes (joints), internodes, and head moves upward. The developing head produces a swollen area or "boot" in the stem. The stem elongation stage ends when the head emerges from the upper or flag leaf sheath. Plants in the boot and tillering stages are attacked by the same types of insects. In addition, stems are vulnerable to attack by the wheat stem sawfly, wheat jointworm, and other stem-boring insects.

Heading

During this stage, the head emerges from the stem, flowering or anthesis occurs, and the kernels begin to fill with "milk." Soft kernels are attacked by several species of aphids, the wheat head armyworm, orange wheat blossom midges, and stink bugs. Plants are

Table 2.1. Major plant growth stages and descriptions

Stage	Description
Seed	From planting to germination
Seedling	Roots and stem appear
Tillering	Secondary stems develop
Stem elongation or boot	Nodes and internodes appear; head moves upward in the stem
Heading	Head emerges from the stem; flowering begins
Ripening	Kernels fill and harden; green coloration is lost; moisture content drops

more tolerant of defoliation after the heading stage is reached. Lower leaves may begin to dry (senesce).

Ripening or Maturation

Kernels harden, green coloration is lost, plant moisture content drops, and senescence occurs during the ripening stage. Developed kernels are attacked by wheat head armyworms and grasshoppers. The leaves and stems are no longer attractive to insect pests.

Other Factors Affecting Insect Activity

Postharvest

The stubble, or lower stems, remain in the field after harvest. Hessian flies, wheat stem sawflies, and wheat jointworms overwinter in the stubble.

Seed that is lost during harvest may germinate in the fall, resulting in volunteer growth. The volunteer growth may provide a "green bridge" between senescence of spring grain and appearance of fall-seeded crops. Volunteer plants are important hosts for aphids and mites that vector plant pathogens.

Hail may damage plants and cause a late-season flush of new plant growth. This new growth supports populations of aphids and mites and provides nutrients that enhance egg production by grasshoppers.

Fig. 2.1. Left to right, grains of wheat, barley, and oats.

Winter and Spring Grain

Winter small grain is planted during the fall. In southern regions, it may be grazed by livestock during the fall or winter. Plants are able to utilize moisture over a longer period than spring grain. Growth resumes in the spring, and fields are harvested during early summer.

Spring grain is planted during the spring, and crop maturity occurs somewhat later than that of winter grain. Spring grain is exposed to insects and diseases for a shorter period than winter grains.

Plant Morphology

Leaves consist of a blade, a sheath that surrounds the stem, a pair of auricles, and a ligule at the junction of the blade and stem. However, the term "leaf" is generally used to refer to the leaf blade. Photosynthesis takes place in leaves, stems, and awns.

Fig. 2.2. Left to right, heads (panicles) of wheat, barley, and oats.

Hardened sections of the stem that give rise to leaf sheaths are called nodes. Internodes are the regions between nodes. Stem-boring insects commonly damage fibrovascular bundles that are constricted at the nodes, which impedes transport of carbohydrates to developing grain.

Crop Identification

Wheat, oats, and barley can be identified in several ways. Seedling plants can be pulled from the ground to recover the seed for identification (Fig. 2.1). Heads (panicles) of mature plants are characteristic for each crop (Fig. 2.2). The junction of leaves and stems can be examined for presence of pubescence, auricles, and ligules (Fig. 2.3).

Plant Health

Healthy, vigorous plants are more tolerant of insect feeding than plants stressed by lack of moisture or nutrients. Therefore, adequate amounts of fertilizer may increase plants' tolerance to insect damage. There is no evidence that well-fertilized crops are more susceptible to insect feeding, nor is there evidence that invading insects select weaker plants.

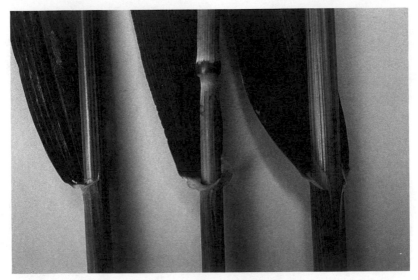

Fig. 2.3. Left to right, stems of wheat, barley, and oats.

Field Plot Problems

The small plot techniques that are commonly used in experimental cultivar trials may produce misleading entomological data. The small plots provide an opportunity for insects to select plants that are in the preferred stage of development or are the tallest. For example, wheat stem maggots select the tallest plants for egg laying. Data would indicate that shorter wheats are resistant. However, the apparent resistance disappears in a large field made up of a single cultivar.

Selected References

Cook, R. J., and Veseth, R. J. 1991. Wheat Health Management. American Phytopathological Society, St. Paul, MN.

Leonard, W. H., and Martin, J. H. 1967. Cereal Crops. Macmillan, New York.

Chapter 3

Insect Biology and Classification

Correct identification of pest insects and an understanding of their biology are important factors in the development and implementation of pest-management programs. Some groups of pest insects produce multiple annual generations, and early populations may indicate forthcoming problems. Crop damage is often caused only by larval insects, and after larval development is finished, there will be no more damage until the following growing season. Therefore, the extent of larval development determined during field inspection is important when estimates of further expected damage are made.

Insect Metamorphosis

Insects go through several life stages. The egg is followed by an immature stage, a pupal stage (in some cases), and finally an adult stage (Table 3.1).

Insects that undergo gradual metamorphosis produce nymphs that are miniature replicas of adults. Nymphs shed their exoskeletons, or skins, between each instar as they grow. They are adults after the final molt. Adult insects usually have wings and are sexually mature. Crickets and grasshoppers are examples of insects with gradual metamorphosis.

Insects with complete metamorphosis undergo major morphological changes between the immature and adult stages. For example, cutworms, or larvae, are very different from the adult moths.

Insects with complete metamorphosis have a pupal stage. This is called a resting stage and usually takes place in a protected location.

Table 3.1. Life stages of insects with gradual or complete metamorphosis

Metamorphosis	Stages
Gradual	egg → nymph → adult
Complete	egg → larva → pupa → adult

Each life stage performs an important function for the insect. The immature stage is a time for feeding and growth. Larvae are very small after emerging from eggs, and they must feed to obtain enough energy to grow and become pupae. Some adults do not feed and must rely upon energy stored during the larval stage. Most crop damage caused by insects is a result of larval feeding, although some beetles (adults) may also be very destructive.

During the pupal stage, larval tissues break down and give rise to new tissues that become the adult insect. This is also known as a resting stage and frequently occurs when conditions are unfavorable for larval development.

During the adult stage, insects mate, select suitable hosts for their progeny, and produce eggs. Egg production makes increases in population size possible. The number of eggs produced by female insects varies greatly among groups. For example, wheat stem sawfly wasps lay only about 35 eggs, but cutworm moths may produce several thousand eggs.

Insect Classification

Insects are the most diverse animal group on earth. Adult insects have three body regions, the head, thorax, and abdomen, and three pairs of legs. Wings may be present or lacking. Spiders and mites are close relatives of insects but have two body regions and four pairs of legs. There are 24 major groups, or orders, of insects. The orders are divided into families that are composed of genera and one or more species. Scientific names are italicized and are followed by the name of the person who initially described the species.

Important species of insects usually also have common names. Acceptable names are published in the Entomological Society of America bulletin titled "Common Names of Insects and Related Organisms." There may also be names for insects that are used by the public but are not used in the literature. An example is "waterbugs" for cockroaches.

Table 3.2. Commonly used names of adult and immature insect life stages

Adult Insect	Immature or Larval Insect
Grasshoppers, crickets, aphids, thrips, and stink bugs	Nymphs
Beetles, weevils, and wasps	Grubs
Moths and butterflies	Cutworms, armyworms, stem borers, and caterpillars
Flies	Maggots

Common names are also used for groups of insects. Fly larvae are called maggots; wasp and beetle larvae are grubs; and moth larvae are called caterpillars, cutworms, and armyworms (Table 3.2).

Correct identification of pest insects is important, and specialists should be consulted for identification of representative specimens. It is beyond the scope of this book to provide the information necessary to identify pests to the species level. However, adequate information is presented to tentatively identify many pests and the damage they cause.

Insect Orders Important in Small Grains

Insects within each order undergo similar metamorphosis and usually have the same type of mouthparts (Table 3.3).

Orthoptera

Insects in the order Orthoptera (grasshoppers) have chewing mouthparts, thickened front wings, membranous hind wings, and enlarged hind legs for climbing and jumping. Important families include the Acrididae (short-horned grasshoppers), Tettigoniidae (long-horned grasshoppers and Mormon crickets), and Gryllidae (field and

Table 3.3. Insects of major importance in small grains

Common Name	Order	Mouthparts	Metamorphosis
Grasshoppers	Orthoptera	Chewing	Gradual
Thrips	Thysanoptera	Rasping	Modified gradual
"True" bugs	Heteroptera (Hemiptera)	Piercing and sucking	Gradual
Aphids	Homoptera	Piercing and sucking	Gradual
Beetles	Coleoptera	Chewing	Complete
Butterflies, moths, and cutworms	Lepidoptera	Chewing (larval stage)	Complete
Flies	Diptera	Rasping	Complete
Bees, wasps, and ants	Hymenoptera	Chewing	Complete

mole crickets). They have gradual metamorphosis (egg → nymph → adult), and there usually is one generation per year.

Thysanoptera

Members of the order Thysanoptera (thrips) are small, fringe-winged insects with rasping mouthparts. The family Thripidae has several species that are of economic importance. They have modified metamorphosis (egg → nymph → prepupa → adult). Some species have one generation per year, and other species have several.

Hemiptera

Insects in the order Hemiptera (Heteroptera), "true" bugs, have piercing and sucking mouthparts that are directed forward, and the front pair of wings folds back flat over hind wings. Some species are predatory and have strong, grasping-type front legs and prominent eyes. Other bugs are phytophagous and feed in the plant phloem.

Important families include the Miridae (plant bugs), Nabidae (damsel bugs), Lygaeidae (lygaeids), and Pentatomidae (stink bugs). These insects have gradual metamorphosis (egg → nymph → adult) and one or more generations per year.

Homoptera

Members of the order Homoptera have piercing and sucking mouthparts, which are directed to the rear, and membranous wings. Important families include the Cicadellidae (leafhoppers) and Aphididae (aphids). They have gradual metamorphosis (egg → nymph → adult) and may have many generations per year.

Coleoptera

Members of the order Coleoptera (beetles) have chewing mouthparts. The front wings are hardened and fold back over the membranous hind wings. Important families include the Carabidae (ground beetles), Scarabaeidae (May beetles and June beetles), Elateridae (click beetles and larval wireworms), Coccinellidae (lady beetles), Tenebrionidae (darkling beetles and false wireworms), and Curculionidae (weevils). They have complete metamorphosis (egg → larva or grub → pupa → adult or beetle). Life cycles may take several months to several years to complete.

Important beneficial groups are found among the Carabidae and Coccinellidae. Economically important pest species appear among the Scarabaeidae, Elateridae, and Tenebrionidae.

Lepidoptera

Insects in the order Lepidoptera (butterflies and moths) have larvae that can be very destructive. The larvae have chewing mouthparts, and the adults have siphoning mouthparts. Important families include the Pyralidae (stem borers), Arctiidae (tiger moths and woolly bear caterpillars), and Noctuidae (cutworms and armyworms). These insects have complete metamorphosis (egg → larva or caterpillar → pupa → adult or butterfly or moth). There may be one or more generations per year, and a single species may have more generations in warm climates.

Diptera

Members of the order Diptera (flies) have one functional pair of wings, and the larvae (maggots) are wormlike and have hooklike mouthparts. Flies have sponging-type mouthparts. Important families include the Cecidomyiidae (gall gnats), Syrphidae (flower flies), Agromyzidae (leaf miners), and Tachinidae (tachinid flies). They have complete metamorphosis (egg → larva or maggot → pupa → adult or fly). Beneficial parasitic or predatory species are found in the Syrphidae and Tachinidae. Crop damage is caused by species in families including the Cecidomyiidae and Agromyzidae.

Hymenoptera

Hymenoptera includes the bees, wasps, and ants. The adults have chewing mouthparts and clear membranous wings. Important families are the Braconidae (braconids), Cephidae (stem sawflies), Eurytomidae (jointworms), Formicidae (ants), Ichneumonidae (ichneumons), Tiphiidae (tiphiid wasps), Trichogrammatidae (trichograma wasps), and Vespidae (hornets and wasps). They have complete metamorphosis (egg → larva → pupa → adult or wasp, bee, or ant). Beneficial species include bees, wasps, ants, and parasitic wasps. Pests are found among the bees, wasps, ants, and Cephidae.

Chapter 4

Insect Trapping and Detection

Monitoring pest insect populations in crops is an essential part of management programs. Sampling methodology ranges from the visual examination of plants to the use of precisely formulated sex attractants called pheromones.

Visual Examination of Plants

Plants can be selected at random and examined to detect aphids and thrips. In some cases, plants can be chosen from areas that are the most likely to be infested, such as field margins, near the borders of damaged areas, or where there is uneven plant growth.

Plant damage or unusual growth symptoms may provide clues for finding pest insects. Small holes in leaves indicate feeding by young cutworms, wireworms, or flea beetles. Thickened, darkened leaves indicate Hessian flies may be present. Aphid feeding produces leaf chlorosis. Plants with white heads may be infested with wheat stem maggots. Stems that are twisted and have swollen nodes may be infested with jointworms. Lodged stems that are filled with "sawdust" and have cleanly cut bases probably have been attacked by wheat stem sawflies. Recent "window pane" feeding indicates the presence of cereal leaf beetle grubs or adults. Many kinds of pest insects are present for short periods of time, but the characteristic damage that they cause provides clues for later diagnosis.

Soil Sampling

Soil-dwelling insects are difficult to detect. Large cutworms and white grubs can be found by searching through the soil. In some cases, plots can be flooded to bring grubs to the surface. Army cutworms are found in the dry upper soil layer and can be collected by gently raking with a small garden rake (Fig. 4.1). Golf hole cutters can be used to collect soil samples of uniform size, and material can be sifted through a series of wire screens to separate insects from stones and fine soil (Fig. 4.2).

Sugar- or salt-water solutions can also be used to separate insects

Fig. 4.1. Raking cutworms from the soil within rows of wheat.

Fig. 4.2. Soil sieve used to collect wireworms from the soil.

from soil. Although insects eggs or larvae normally are heavier than water and sink to the bottom of containers, the specific gravity of the water can be changed by adding sugar or salt. Insects will then float to the surface, where they can be collected and separated from soil and heavy debris.

Solar bait station technology was developed for use in corn in

Fig. 4.3. Solar bait station used to attract wireworms.

Fig. 4.4. Sweep net used to collect foliage-dwelling insects.

Kansas and is useful for detection of insects in small grains. However, because spring wheat is planted much earlier than corn, traps must be installed in the fall and examined prior to spring planting. Stations are constructed by burying a small handful of grain about 10 cm below the soil surface. A mound of soil about 20 cm high is built to increase the surface area that is exposed to the sun. The station is covered with a piece of clear plastic that is 90 cm square and 8 mm or more thick. The edges of the plastic are covered with soil and marked with a surveyor's flag so the station can be found in the spring (Fig. 4.3). During evaluation, the bait and soil in the mound are examined to detect wireworms. Insects are attracted to the warm, moist soil under the plastic and remain to feed on the bait. The grain, or bait, may release carbon dioxide that attracts insects. It is easier to find insects that are concentrated under the stations than to search randomly in the soil.

Fig. 4.5. Walk-in light trap used to collect night-flying insects.

Traps

Sweep Nets

Sweep nets (Fig. 4.4) are effective for collecting foliage-dwelling insects such as wheat stem sawflies, aphids, wheat stem maggot adults, and many kinds of insect predators. Sweep net sampling is comparatively cheap and easy and can begin as soon as plants emerge from the soil and continue until mature plants become dry and brittle. Heavy-duty sweep nets are required for plant sampling.

Light Traps

Night-flying insects such as moths and June beetles can be collected in light traps. Trap styles vary from large, walk-in cages (Fig. 4.5) to small, portable, battery-operated units (Fig. 4.6). Catch sizes are affected by changes in lunar light intensity and competition from nearby lights. Both male and female insects are captured. Many

Fig. 4.6. Portable, battery-operated light trap used to collect armyworm and cutworm moths.

Fig. 4.7. Suction trap for collecting migrating aphids.

Fig. 4.8. Sticky trap used to collect Hessian fly adults and other small insects.

kinds of insects are captured in light traps, and it is necessary to sort through samples to find the target species. Light sources include mercury vapor, ultraviolet, and fluorescent bulbs. Wavelength of emitted light varies with these sources and influences attractiveness to insects.

Aerial Traps

Tall suction traps can be used to capture airborne, dispersing aphids (Fig. 4.7). Also, aerial nets pulled by airplanes have been used for high elevation sampling. Nets can also be attached to cars or to the ends of motorized rotating arms for elevated stationary traps.

Sticky Traps

Commercial formulations of petroleum-based sticky material can be used to capture Hessian flies and other small flying insects. The sticky material can be applied to colored or white cardboard that is

Fig. 4.9. Pitfall trap used to collect insects that are active on the soil surface.

Fig. 4.10. Top, pitfall trap constructed from a funnel and plastic jar; bottom, cross section of a pitfall trap showing the wire screen base that permits drainage of rainfall.

Fig. 4.11. Emergence trap used to collect emerging soil-inhabiting insects.

Fig. 4.12. Emergence trap with four support legs and an aluminum angel food cake pan and upper cage to capture insects.

Fig. 4.13. Top, emergence trap cone and upper cage supported by an internal stake; bottom, captured insects are collected through the base of the upper cage.

stapled to wooden stakes (Fig. 4.8). Individual traps are effective in the field for about 1 week or until they are covered with blowing debris.

Yellow Pan Traps

Flat-bottomed pans supported on stakes are filled with water and a small amount of detergent. Pans can be painted yellow to attract aphids and other insects. Insects drown in the water and are retrieved from the bottom.

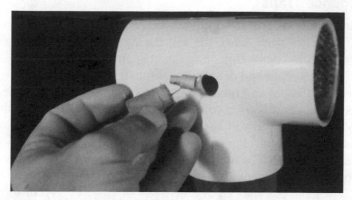

Fig. 4.14. A rubber septum is impregnated with cutworm sex pheromones; the septum is then pinned to a rubber stopper and inserted into the trap.

Fig. 4.15. Cutworm pheromone trap constructed from a polyvinyl chloride pipe joint. Moths enter through wire screens at the upper openings and fall into the removable plastic jar at the bottom.

Pitfall Traps

Insects such as ground beetles and false wireworms that are active on the soil surface can be captured in pitfall traps (Figs. 4.9 and 4.10). Traps can be constructed with catch containers filled with a solution of ethylene glycol, water, and soap, or the catch containers can have screen bottoms to permit rain water to flow through. Catch sizes can be increased by using strips of metal to guide insects toward the traps. Catch sizes are indicators of "activity density," the number of insects present and their activity.

Emergence Traps

Many kinds of soil-dwelling insects, such as wireworms, white grubs, and cutworms, leave the soil during the adult stage to mate, disperse, and lay eggs. Emerging insects can be captured to estimate population densities and determine seasonal activity.

Various kinds of emergence traps can be constructed. A style that has been successfully used consists of a screen collection cone that is tightly secured to the soil surface. Cones can be supported with an internal stake (Fig. 4.11) or by an outside frame consisting of a platform and four legs. The upper chamber may consist of an angel food cake pan that surrounds the upper end of the collection cone (Fig. 4.12). The cake pan may be dry, or water or ethylene glycol solution may be added to immobilize insects.

Fig. 4.16. Drop cup pheromone trap for collecting cutworm moths.

A simpler trap can be made with an aluminum cone that supports a removable cage consisting of wire screen and a plastic petri dish top and bottom (Fig. 4.13).

Pheromone Traps

Pheromone traps are effective for estimating flight activity of moths. Data are useful in predicting the potential for subsequent cutworm outbreaks. Sex pheromones consist of several chemicals in precise ratios. For trapping purposes, rubber septa are impregnated with the necessary chemicals (Fig. 4.14). Pheromones are species specific, and catches usually do not need to be sorted.

Several styles of traps can be made. One style consists of a T-shaped plastic pipe. Wire screen funnels are fitted into the upper pipe ends, and a removable catch container is pressed into the lower section (Fig. 4.15). This design permits the circulation necessary for dispersal of pheromones and also protects the captured insects from

Fig. 4.17. Sticky trap for collecting cutworm moths.

Fig. 4.18. Berlese funnel for extracting small insects from soil and plant debris.

rain. Traps can be suspended from fence wire (Fig. 4.16) or fastened to posts. The polyvinyl chloride pipe fittings are available in various sizes. Small sizes are acceptable if traps are serviced weekly, but larger sizes should be used if traps are to be serviced monthly.

Several styles of sticky traps designed for pheromone attractants are available from commercial sources (Fig. 4.17). However, sticky surfaces are covered with moths during heavy flights, and catch sizes may not accurately reflect flight sizes.

Berlese Funnels

Small, reclusive insects can be separated from plant material in Berlese funnels. Electric lamps are suspended over field-collected debris, and the heat and light drive insects downward into alcohol containers (Fig. 4.18).

Selected References

Kogan, M., and Herzog, D. C. 1980. Sampling Methods in Soybean Entomology. Springer-Verlag, New York.

Lester, D. G., and Morrill, W. L. 1989. Activity density of ground beetles in alfalfa and sainfoin. J. Agric. Entomol. 6:71-76.

Morrill, W. L. 1975. Plastic pitfall trap. Environ. Entomol. 4:596.

Morrill, W. L. 1978. Emergence of click beetles (Coleoptera: Elateridae) from some Georgia grasslands. Environ. Entomol. 7:895-896.

Morrill, W. L. 1984. Wireworms: Control, sampling methodology, and effect on wheat yield in Montana. J. Ga. Entomol. Soc. 19:67-71.

Morrill, W. L. 1988. Evaluation of pheromone trap designs. J. Econ. Entomol. 81:735-737.

Morrill, W. L., Lester, D. G., and Wrona, A. E. 1990. Factors affecting efficacy of pitfall traps for beetles (Coleoptera: Carabidae and Tenebrionidae). J. Entomol. Sci. 25:284-293.

Southwood, T. R. E. 1987. Ecological Methods. Chapman and Hall, London.

Chapter 5

Methods of Loss Assessment

Research trials designed to determine the relationship between pest insect population densities and crop yields provide a portion of the data needed to establish economic thresholds. However, loss assessment trials are difficult to manage because of variations in soil, moisture stress, plant vigor, and other uncontrolled factors. The following techniques have proved to be effective.

Natural Infestations

Although naturally occurring pest insect populations can cause obvious reductions in crop yields, measurements of loss may be impossible because there are no uninfested plants for comparison. In some cases, plant damage can be measured in locations in which there are different pest population densities within the same field.

For example, an uneven infestation of wireworms in a winter wheat field resulted in obvious differences in plant stand and vigor. To document the relationship between insect density and plant damage, regions of the field were visually rated "very thin," "thin," and "normal" on the basis of the appearance of the crop. Data were then collected from five areas within each of these categories. The soil

Table 5.1. Effect of various wireworm population densities on winter wheat

Field Appearance	Healthy Plants per Meter	Dead Plants per Meter	Damaged Plants per Meter	Wireworms per Meter
Very thin	11.0	4.5	6.5	10.0
Thin	16.0	2.0	4.5	2.5
Normal	24.0	0	0	0

within the rows was carefully examined, and the numbers of plants that were dead, damaged, or healthy and the numbers of wireworms were recorded (Table 5.1). Plants could have been harvested from each area to determine ultimate effect on yield.

Table 5.2. Effect of Hessian fly infestation on wheat attacked during the stem elongation stage[a]

	Kernels per Head	Head Weight (g)
Uninfested	22.8 a	91.0 a
Infested	15.8 b	38.1 b

[a] Numbers within columns followed by the same letter are not significantly different by ANOVA (LSD, $P < 0.05$).

Table 5.3. Effect of wheat jointworms and sheathminers on wheat head weight[a]

Plant Infestation	Head Weight (g)
Jointworm	0.3 a
Jointworm and sheathminer	0.4 a
Uninfested	1.0 b
One sheathminer	1.3 bc
Two sheathminers	1.5 c

[a] Means followed by the same letter are not significantly different by ANOVA (LSD, $P < 0.05$); $n = 200$ plants.

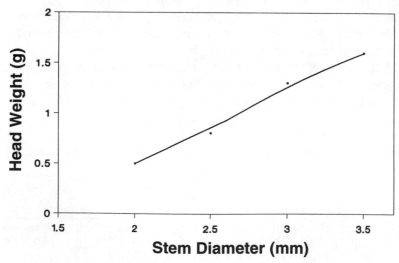

Fig. 5.1. Relationship between wheat stem diameter (plant size) and head weight.

In some cases, yield loss can be estimated by comparing head weights of individual infested and uninfested plants, as in the case of wheat infested by the Hessian fly (Table 5.2). However, data from individual plants may indicate that the largest stems are more likely

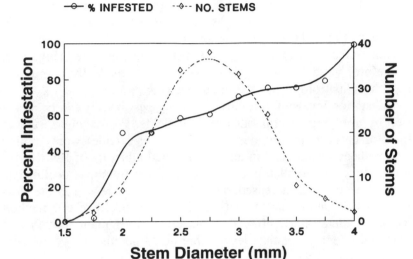

Fig. 5.2. Frequency distribution of wheat stem size and percent infestation by wheat stem sawflies in stems of various sizes.

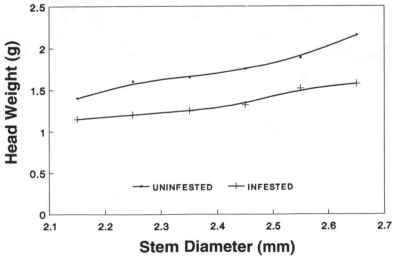

Fig. 5.3. Comparison of head weight of wheat plants of similar size and stems uninfested and infested with wheat stem sawflies.

to be infested by some pest insects. For example, sheathminers pre-fer plants that produce the heaviest heads (Table 5.3). Therefore, plants may need to be categorized according to size so that precise comparisons can be made.

Plant-size categories can be based on stem diameter measured at a preselected point, such as midway between the second and third nodes from the top. Head weight increases with an increase in stem diameter (Fig. 5.1). In the example of the wheat stem sawfly, the percent infestation increases as stem diameter increases (Fig. 5.2). Therefore, valid estimations of loss must be made by comparing head weights from plants of similar size (Fig. 5.3).

Infestation levels of naturally occurring pest insects can be modi-fied by applying several rates of insecticide. For example, several rates of carbofuran granules (a systemic insecticide) were applied when wheat was planted in an area that had a history of Hessian fly infestation. The population density of Hessian fly pupae declined as the carbofuran rate increased, and wheat yield increased as the Hes-sian fly infestation decreased (Fig. 5.4). Drawbacks of this method include undetermined effects of insecticides on plant growth and potential effects on nematodes or other nontarget insects.

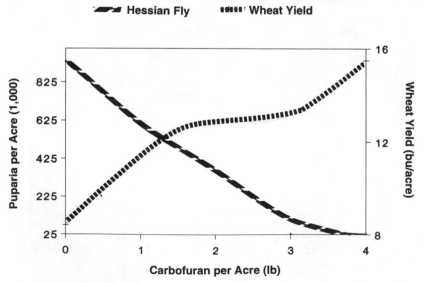

Fig. 5.4. Relationship among Hessian fly infestation level, wheat yield, and pesticide application rate.

Table 5.4. Survival of slender ricebugs and weight of rice panicles after 14 days of confinement in cages constructed of various materials[a]

Cage Material	Bug Survival (%)	Panicle Weight (g)	
		No Bugs	Four Bugs
Clear plastic	98.1 a	3.1 a	1.3 a
Nylon cloth	93.5 a	3.0 a	1.5 a
Wire screen	93.5 a	2.7 a	2.1 a
Uncaged panicles	...	2.6 a	...

[a] Means followed by the same letter within columns are not significantly different by ANOVA (LSD, $P < 0.05$); $n = 32$ panicles.

Simulation

Leaves can be removed from plants to simulate defoliation damage caused by insects, and heads can be harvested to estimate losses. Some variables that can be tested include damage that occurs during different plant growth stages and different amounts of leaf area lost. Difficulties with this method include estimation of the amount of feeding by the target insect and differences in plant response to insect feeding and artificial defoliation.

Caging

Cages can be used to restrain pest insects on host plants. Trials can be designed to determine effects of many variables such as insect density and susceptibility of plants during various growth stages.

Fig. 5.5. Cloth cages supported by bamboo stakes and used to protect rice from slender ricebugs in the Philippines.

Cages affect the extent of exposure of plants to sunlight, wind, and rainfall. Therefore, the use of cages should be minimized, and effects of cages on plants must be measured. Treatments in a typical experiment would consist of uncaged plants, caged plants without

Fig. 5.6. Aluminum wire screen cages used to confine pest insects on individual plants.

Fig. 5.7. Insects can be introduced through holes in cage tops.

insects, and caged plants with insects. Insect mortality in the cages during the test period should be monitored (Table 5.4).

Several kinds of cages can be successfully used in small grain loss assessment trials. Fairly large areas can be caged with wooden posts and frames covered with wire or fiberglass screen (Fig. 5.5). Commercial cages are available.

Single plants may be covered with cages made from a polyvinyl chloride pipe 10 cm long and 34 cm in diameter pressed into the ground to form the base of a cylinder of aluminum wire screen of adequate height to enclose the plant (Fig. 5.6). The top of the cage can be covered with a petri dish (Fig. 5.7). Cages can be supported with wooden stakes that are driven into the ground.

Cages for individual heads can be made from plastic cups of vari-

Fig. 5.8. Enclosing individual heads of barley in cages made from plastic jars.

Table 5.5. Effect of stink bug feeding on head weight, number of kernels, and kernel weight[a]

Bugs per Head	Head Weight (g)	Kernels per Head	100-Kernel Weight
Uncaged			
0	1.5 a	44 a	3.0 a
Caged[b]			
0	1.7 a	53 a	3.2 a
1	0.9 b	30 b	2.9 ab
2	0.8 b	31 b	2.5 bc
4	0.4 b	20 b	2.1 c

[a] Means within columns followed by the same letter are not significantly different by ANOVA (LSD, $P < 0.05$); $n = 25$ heads.
[b] Bugs were eaged on individual heads from early heading through plant maturity.

ous sizes. The bottom of each cup is removed, and fine cloth is glued around the perimeter. The cup is lowered over a plant head, and the cloth base is then tied around the stem to prevent escape of the insects (Fig. 5.8). Holes can be cut in the cup lids to provide air circulation and reduce growth of mold on the heads. Cages can be supported with cane stakes.

The concept of "insect-days" is useful for estimating effects of the duration and intensity of insect feeding. The duration of insect feeding must be separated from the intensity. For example, 10 insect-days can represent 10 insects feeding for 1 day and also one insect feeding for 10 days. Plants are less tolerant of heavy damage that occurs over a short period of time than of light damage that occurs over a long period. Therefore, "feeding intensity" is used in analysis of insect-day data.

Evaluation

Evaluations can include effects on plant height, total head weight, weight of grain, percent protein, weight per 100 kernels, and estimates of kernel damage. Total head weight is the easiest estimate to obtain because kernels need not be threshed and cleaned. More differences appeared when kernel weights were compared than when total head weights and number of kernels per head were compared in a stink bug trial (Table 5.5).

Variations in grain production among plants is affected by several factors. The primary stem appears earlier and matures sooner than stems of secondary tillers. The head and stem of a primary stem are also heavier. The ability of plants to compensate for lost stems or heads is complicated and may be affected by plant age and availability of moisture and nutrients.

Selected References

Bardner, R., and Fletcher, K. E. 1974. Insect infestations and their effect on the growth and yield of field crops: A review. Bull. Entomol. Res. 64:141-160.

Capinera, J. L., and Roltsch, W. J. 1980. Response of wheat seedlings to actual and simulated migratory grasshopper defoliation. J. Econ. Entomol. 73:258-261.

Poston, F. L., Pedigo, L. P., and Welch, S. M. 1983. Economic injury levels: Reality and practicality. Bull. Entomol. Soc. Am. 29:49-53.

Ring, D. R., Benedict, J. H., Landivar, J. A., and Eddleman, B. R. 1993. Economic injury levels and development and application of response surfaces relating insect injury, normalized yield, and plant physiological age. Environ. Entomol. 22:273-282.

Walker, P. T. 1983. The assessment of crop losses in cereals. Insect Sci. Appl. 4:97-104.

Weiss, M. J. 1987. Influence of simulated grasshopper damage on yield and quality components of spring-planted wheat, barley, and oats. J. Kans. Entomol. Soc. 60:77-82.

Chapter 6

Damage Caused by Insects

Yield and quality of small grains are reduced by many biotic factors including insects, weeds, and plant pathogens. Decreasing profit margins and increased awareness of potential environmental hazards from pesticide use have increased the need for accurate crop loss estimates.

The amount of damage caused by pest insects in small grains is influenced by many factors. Some plant growth stages, especially the seedling and flowering stages, are sensitive to defoliation. Stress caused by lack of available moisture or nutrients reduces the ability of the plant to compensate for insect feeding. Also, the type of plant damage determines the effect on host plants (Table 6.1).

Defoliation

Insects such as grasshoppers and cutworms that have chewing-type mouthparts feed on foliage, reducing the amount of leaf area available for photosynthesis. Damage to the leaf surface (for example, that done by cereal leaf beetles and thrips) also reduces the amount of effective photosynthetic area.

Root Damage

Plant roots are attacked by white grubs and wireworms. These insects are less obvious and more difficult to detect than foliage feeders, but they can cause significant crop damage. Plant wounds caused by insect feeding can provide entry sites for plant pathogens.

Table 6.1. Some important kinds of damage caused by pest insects in small grains

Insect Damage	Effect on Host Plant
Defoliation	Reduced photosynthetic area, poor growth or death, secondary infection by pathogens
Stem boring	Weakened plants, poor head development
Phloem feeding	Reduced plant growth or vigor, stunting from feeding toxins, transmission of plant pathogens
Specialized damage	
Gall formation	Poor head formation
Stem cutting	Lodged plants
Root feeding	Weakened plants, poor growth or death, secondary infection by pathogens

Specialized Feeding

Some insects with chewing mouthparts cause plant damage greater than that of simple defoliation. Pale western cutworms cut plant stems near the soil surface, and the damaged plants fall to the ground. The plants may be only partially consumed even though the entire plant is destroyed. Grasshoppers may be attracted to the upper portion of the stem that supports the plant head because this is the last part of the plant that remains green. When stems are severed, heads fall to the ground and are lost. Stem-boring caterpillars and sawflies damage vascular tissue in stems, and affected plants are unable to transfer nutrients through nodes to developing kernels. Wheat head armyworms feed on kernels, and because damaged grain is light in weight, it may pass through combines during harvest.

Feeding on Plant Fluid

Aphids, stink bugs, and other plant bugs have piercing and sucking mouthparts that are used for feeding in cells, phloem, and developing kernels. Affected plants may lack vigor or produce shrunken, low-quality grain. Also, aphids release large amounts of sticky fluid (honeydew) from the anus. This fluid supports growth of mold on plant surfaces, and photosynthesis is inhibited. Sticky leaves may fail to unfold properly, and emerging heads may be distorted or "goose-necked."

Feeding Toxins

Some species of aphids and Hessian fly maggots inject toxic saliva into host plants. Affected plants have white or yellow streaks in the leaves, weakened stems, and dead tillers.

Vectoring of Plant Pathogens

Plant pathogens may be carried from diseased to healthy plants by winged insects or windblown arthropods. Most of these insects have piercing and sucking mouthparts. For example, the causal agent of barley yellow dwarf is carried by aphids.

Plant Modification

Phytophagous insects sometimes enhance their chances of survival by physically modifying host plants. For example, plants develop galls in response to jointworm toxin. Gall tissue provides nutrition and protection for the jointworm grubs. Grain production by infested stems is greatly reduced. Although there is no equivalent in small grain in North America, the Asian rice leaffolder laces plant leaves together to make a protective retreat. The area of leaves exposed to the sun is reduced, and crop loss occurs.

Several kinds of insect larvae modify host plants to permit escape of the fragile adult. Wheat stem sawfly wasps cannot chew through stem walls of host plants. Stems are therefore severed by larvae, and emerging wasps exit the plants by pushing through soft plugs left by larvae at the top of the stubble.

Wheat stem maggots make circular cuts around the upper plant stems. The stems desiccate, shrink, and loosen within the leaf sheath. Newly emerged flies are then able to escape from the damaged plants.

Hessian fly larvae inject toxin into plants. Leaf sheaths loosen and provide an escape route for the fragile, newly emerged flies.

Evolution

There has been natural selection for survival of phytophagous insects that are efficient feeders or that have minimum impact on the survival and competitiveness of their host plants. Enhanced plant damage, i.e., damage in addition to that necessary for obtaining insects' food for growth, usually can be explained by enhanced larval protection or provision for adult escape from plants. Selection for development of aphid feeding toxins is more difficult to interpret.

Chapter 7

Insect Control

Insect outbreaks can appear suddenly and, if untreated, may completely destroy crops. High numbers of adult grasshoppers can quickly disperse from rangeland to small grain fields. Cutworms rapidly develop from eggs, and crop foliage can disappear almost overnight. Early detection of these sporadically occurring insects is important. Applied control practices must quickly suppress the target organisms to prevent economically important crop losses.

Some insect pests cause chronic problems, and prophylactic measures can be effective in reducing crop losses. For example, resistant cultivars can be selected for planting in regions where there is a history of damage by Hessian flies and wheat stem sawflies. Seed can be treated for use in fields where wireworm damage has occurred.

The comparatively low profit margin in small grain production is a limiting factor in selection of control practices. Control methods include insecticides, resistant cultivars, cultural practices, and biological control (Table 7.1).

Chemical Control

Chemicals that are used to kill insects are called insecticides. Insecticides are available in many formulations, the most commonly used of which are emulsifiable concentrates and granules. Systemic insecticides are absorbed into plants and are effective against aphids and other insects that suck plant juices. Systemic insecticides are also applied to control insects that are concealed when they bore into plants. Insecticide registrations and recommendations change periodically, and recent information is available from appropriate government offices.

Table 7.1. Some effective methods of controlling insect pests in small grains

Method	Example
Insecticides	
Contact	Malathion
Systemic	Carbofuran
Host plant resistance	
Avoidance	Early or late growth
Tolerance	Plant regrowth
Physical characters	Pubescence and solid stems
Antibiosis	Plant toxins that kill larvae

People applying insecticides must be familiar with label restrictions, and care should be taken not to exceed the preharvest intervals and application rates. Regulations concerning safe storage, application, and container disposal must be understood and met. The least toxic materials should be used when possible.

Disadvantages of chemical controls include possible negative effects on wildlife, beneficial insects, and groundwater. Repeated insecticide application may result in the development of insect resistance. However, these problems can be prevented by proper selection and application of pesticides.

Host Plant Resistance

Selection of resistant cultivars is important in pest insect management in small grains. Resistant cultivars are available for Hessian flies, greenbugs, and wheat stem sawflies. Resistance may be in the form of avoidance (plant may mature before pests appear), tolerance (plant is able to withstand or recover from attack), physical characteristics (solid stems and pubescence), and antibiosis (plant produces toxins that kill or inhibit insect growth). Some resistant cultivars have lower yield potential and greater susceptibility to lodging. Also, insects may overcome plant resistance characteristics through natural selection in the same fashion that resistance to insecticides occurs.

Cultural Controls

Delayed planting of winter wheat to avoid the fall generation of Hessian flies is an effective cultural control method for some locations. Delayed fall planting also reduces Russian wheat aphid populations but results in later maturation of wheat in the spring and greater potential for wheat stem sawfly damage.

Effective weed control in crops reduces attractiveness to ovipositing insects and eliminates flowers that provide nectar for pest insects. Clean summer fallow eliminates weed hosts for wireworms.

Tilling stubble affects the overwinter survival of wheat stem sawflies, Hessian flies, and wheat jointworms. However, the parasitoids and predators that also may overwinter in stubble probably are affected as well.

Wild grasses in road ditches, fence rows, ravines, and other nonproductive areas are sources of pest insects including stink bugs, wheat stem sawflies, and wireworms. In some cases, grass can be mowed and eliminated as a host for stem-boring insects and head-feeding insects.

Crop rotation is an important cultural control method. In areas of adequate rainfall, soybeans, peanuts, alfalfa, and other nongrass crops can be grown, and populations of pest insects that are obligate grass feeders are reduced.

Approved agronomic practices, including adequate fertilization and irrigation, increase the ability of plants to tolerate insect feeding.

Biological Control

Few applied biological control practices have been developed for controlling pest insects in small grains. Passive control, i.e., the activity of naturally occurring native parasites and predators, is important in suppressing pest populations. Augmentation, the routine mass rearing and releasing of parasitoids, has not been a feasible control method for small grain insect pests. Conservation practices to enhance populations of beneficial biological organisms are not compatible with production practices.

Classical biological control, the introduction of predators or parasitoids for exotic pest species, has sometimes been successful. For example, parasites of the introduced cereal leaf beetle were collected in Europe and released in the United States.

Biological pesticides that are commercially available include *Nosema locustae* for grasshoppers and *Bacillus thuringiensis* var. *kurstaki* for cutworms. Use of these materials is hindered by comparatively high cost, slow action, and inconsistent efficacy.

Pest Control Programs

Insect pest outbreaks may originate from foreign introductions or from changes in population densities of native pests. Factors that affect native pest populations include variations in climate, decreases in parasitoid prevalence, and changes in cropping practices.

Development of management programs in response to pest outbreaks can proceed in an orderly fashion (Fig. 7.1). Reports of new outbreaks usually are in response to crop damage and not to the appearance of an unusually high number of insects. These reports of damage commonly originate from growers who contact local authorities about the problem. Therefore, the first step is to identify the insects responsible for the crop damage. Specimens must be properly identified to determine whether new pests have appeared, and correct names are necessary for literature reviews.

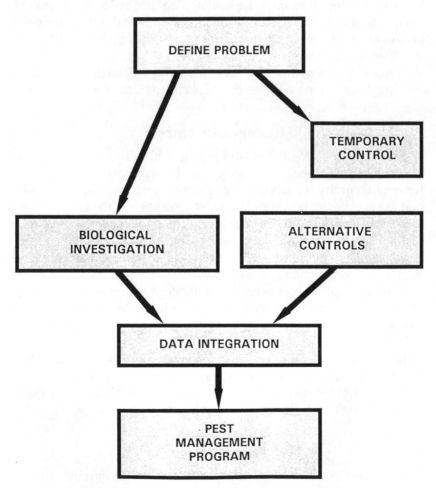

Fig. 7.1. Steps in the development of an insect pest management program.

After the pest has been identified, short-term control practices need to be quickly implemented to minimize crop damage. Insecticides usually are used in crisis situations and may be utilized even if the target pests are not listed in the directions for use of the insecticide. Short-term or emergency controls protect crops and provide time for development of management programs.

The relevant literature should be reviewed to determine the current status of knowledge of the biology, interaction with host plants, control strategies, and origin of the target insect.

Data gaps for control, biology, and behavior should be filled by conducting relevant research. Control strategies, including development of resistant cultivars, modification of cultural practices, and other options, should be field tested. Safer insecticides, minimum rates, and timing of applications to minimize effects on beneficial insects might need to be developed. The existing parasitoid and predator complex should be determined, and in the case of introduced pests, consideration should be given to introductions of exotic beneficial insects. Crop loss data may be needed to establish thresholds. Standardized methodology for sampling and monitoring pest populations should be developed.

The resulting management program must integrate the biological information, current crop values, and potential effects on the environment. Flexibility must be maintained to permit periodic refinement of the program.

A successful program must reduce crop losses and be compatible with production practices, cost effective, and environmentally sound.

Integrated Pest Management

Integrated pest management (IPM) in itself is not a control method but represents selection of the best available management practices. These practices should be implemented only when economic thresholds are expected to be exceeded.

An economic threshold is a pest population density that represents a future population in which the value of the crop loss exceeds the cost of control. The point at which the control cost and crop loss are equal is known as the economic injury level. Treatments are applied to prevent pests from reaching this level.

Insect management differs from weed control because weed control practices at low population densities are justified to prevent later outbreaks. For example, consultants do not recommend treatment of noneconomic levels of cutworms to prevent heavy infestations during the following year.

Insect populations are also more dynamic than weed populations, and significant changes in densities occur in response to biotic and abiotic factors. Many tons of grasshopper bait were applied during the outbreaks of the 1930s, but the end of the outbreak was the result of an increase in precipitation rather than effects of the bait.

Scouting is an important factor in IPM. Producers of small grains are less likely to pay for scouts because of the comparatively low crop value. As a result, many pest infestations in small grains currently are undetected. Enhanced IPM in small grains would result in more scouting, better detection of pest infestations, and ultimately an increase in insecticide use. The low profit margin of small grains also discourages unnecessary application of insecticides.

Selected References

Onstad, D. W. 1987. Calculation of economic-injury levels and economic thresholds for pest management. J. Econ. Entomol. 80:297-303.

Pedigo, L. P., Hutchins, S. H., and Higley, L. G. 1986. Economic injury levels in theory and practice. Annu. Rev. Entomol. 31:341-368.

Poston, F. L., Pedigo, L. P., and Welch, S. M. 1983. Economic injury levels: Reality and practicality. Bull. Entomol. Soc. Am. 29:49-53.

Stern, V. M. 1973. Economic thresholds. Annu. Rev. Entomol. 18:259-280.

Chapter 8

Hail Damage

Damage to plants caused by hail is sometimes confused with damage caused by insects. Correct damage diagnosis is important when fields are evaluated by representatives of hail insurance companies. Some commonly occurring plant damage symptoms caused by hail or freezing are discussed in this section.

White, or "blasted," heads are caused by hail and also by wheat stem maggots (Chapter 22). To determine the cause of the injury, the white heads and associated stems should be pulled from the plant. Ragged, chewed stem ends indicate maggot damage, and cleanly cut stems indicate hail damage (Plate 1). Also, hail damage causes external pock marks on stems (Plate 2).

Twisted, "goosenecked" stems are caused by heavy aphid infestation during the boot stage (Chapter 14) and also by hail during stem elongation (Plate 3). These symptoms may not appear for several days after the hail occurred. Heavy aphid infestations leave a sticky honeydew residue and commonly attract lady beetles. The appearance of lady beetles, larvae, and/or pupae indicates that aphids have been present.

Broken or lodged stems are caused by hail and several kinds of pests. Stems cut by wheat stem sawflies are cleanly broken and filled with sawdust (Chapter 19). Cutworms partially consume plants, and broken ends of lodged plants are ragged. Stems weakened by Hessian flies will contain "flaxseeds." Ground squirrels or gophers feed on plants, and damage is more extensive in the vicinity of their burrows. Grazing by deer and antelope can be identified by droppings and tracks.

Shattered heads are caused by grasshoppers and wheat head armyworms feeding on developing heads. Hail may also break heads or dislodge kernels from sections of heads (Plate 4).

Hail causes ragged, broken leaves or loss of leaves. In contrast, leaf damage from insect feeding includes transparent streaks by cereal leaf beetles, ragged leaf margins by grasshoppers and cutworms, and light-colored streaks by thrips.

Chapter 9

Grasshoppers
(Orthoptera: Acrididae)

Some Important Species

There are about 600 species of grasshoppers in the United States and Canada, but most of the crop damage is caused by five species: the migratory grasshopper, *Melanoplus sanguinipes* (Fabricius); the differential grasshopper, *M. differentialis* (Thomas); the twostriped grasshopper, *M. bivittatus* (Say); the clearwinged grasshopper, *Camnula pellucida* (Scudder); and the bigheaded grasshopper, *Aulocara ellioti* (Thomas).

General Characteristics

Description

Grasshoppers have enlarged hind legs that are used for jumping. A pair of long, narrow front wings extends backward over the abdomen and covers a pair of folding hind wings. Nymphs and adults have chewing mouthparts (Plate 5).

Many species of grasshoppers occur in the United States and Canada. Experts should be consulted for positive identification. However, there are some general guidelines to important groups.

Long-Horned Grasshoppers and Katydids

The antennae are longer than the body (Plate 6).

Grasshoppers

Short-horned grasshoppers: antennae are shorter than the body; this group includes most economically important species.

Spur-throated grasshoppers: a knob can be seen between the front legs (Plate 7).

Banded winged grasshoppers: hind wings are usually brightly colored; noisy flyers.

Slant-faced grasshoppers: top of head points forward.

Distribution

Grasshoppers are found throughout semiarid regions of the United States and Canada. Population densities usually are higher west of the 100th meridian.

Life Cycle

Grasshoppers undergo gradual metamorphosis. Life stages are the egg, nymph, and adult. Eggs of most economically important species overwinter in the soil in pods (Plate 8).

Embryo development begins with rising soil temperatures in the spring. Nymphs emerge over a period of several weeks, but emergence may be delayed by cool weather.

Newly emerged nymphs are about 5 mm long and resemble adults (Plate 9). They have chewing mouthparts and feed on succulent spring vegetation. The exoskeleton, or outer skin, is shed four or five times during development. The adult stage is reached after 35–50 days. Adult grasshoppers have fully developed wings and are sexually mature. There usually is only one generation per year in northern regions.

Grasshoppers mate and lay eggs over a 3-month period. Each female may produce 200–400 eggs. Rangeland, road ditches, and other grassy areas are preferred oviposition sites, although some eggs are laid in cultivated cropland. A plentiful food supply encourages maximum egg production.

High population densities of grasshoppers are enhanced by early, warm springs that favor low mortality of nymphs and by extended egg-laying periods in the fall resulting from delayed occurrence of the first killing frost. Overwinter egg survival rates are high when moderate moisture and adequate snow cover are present.

Detection, Sampling, and Monitoring

Grasshopper abundance can be estimated by counting the number of adults observed while walking 100 m through fields or along roadsides. Areas with high levels of adult activity can later be sampled to determine the prevalence of egg pods. These areas can also be monitored in the spring to determine activity of nymphs.

Crop Damage and Losses

Feeding damage by grasshoppers in early summer in small grains includes defoliation (Plate 10), head damage (Plate 11), and stem damage (Plate 12). Weakened stems may lodge, resulting in harvest losses. Winter wheat may also be defoliated in the fall, and damage is frequently more severe along field borders (Plate 13).

Outbreaks usually follow several years of favorable conditions and gradual population increases. Annual four- to 10-fold increases in population may occur. Small grain crops may be completely destroyed unless control measures are taken.

Cropland infestations usually result from population increases in rangeland. Grasshoppers are readily visible and therefore are more likely to be detected than many other kinds of insect pests.

Grasshopper outbreaks have appeared at irregular intervals. Heavy infestations occurred in the northern Great Plains during drought periods of 1917–1918, 1921–1925, and 1934–1941. Dry conditions

Fig. 9.1. Neighborhood mixing bee at which grasshopper baits were formulated.

suppress pathogen activity, and drought-stressed crops are less tolerant of defoliation.

Commercial grasshopper baits were not available during early grasshopper outbreaks. Growers worked together to make their own bait, consisting of sodium arsenite or sodium fluorosilicate and wheat bran, in neighborhood "mixing bees" (Figs. 9.1 and 9.2). The baits and dusts were applied by hand (Fig. 9.3), from horse-drawn equipment (Fig. 9.4), from automobiles (Fig. 9.5), and finally by aircraft (Fig. 9.6). Control programs in Montana saved crops valued at $10.6 million during 1934–1936, with a return of $10 for each dollar spent for control.

Fig. 9.2. Grinding lemons to use as an attractant in grasshopper bait.

Fig. 9.3. Applying insecticide dust for grasshopper control.

Fig. 9.4. Horse-drawn equipment applying insecticide dust for grasshopper control.

Fig. 9.5. Distributing grasshopper bait by hand from an automobile.

Fig. 9.6. Aircraft used to distribute grasshopper bait.

Early control practices also included the use of "hopper dozers" (Fig. 9.7). Effectiveness was not documented, but it was recorded that the grasshoppers were dried and used for poultry food.

Economic Thresholds

Economically important crop loss can be expected if egg pod densities exceed 10 per square meter in cropland or 25 per square meter in field borders. Nymph infestations of 12–25 per square meter or adult infestations of 10 per square meter are of economic importance. Severe damage may occur if more than 17 or 9 adults are seen in 100 m along roadsides or in cropland, respectively.

Control

Biological Control

Several pathogens, including *Entomophthora grylii* Fresenius and *Beauveria bassiana* (Balsamo), provide passive biological control. Pathogens are more effective under warm, humid conditions. Fungal

Fig. 9.7. Grasshoppers captured with hopper dozers were dried, bagged, and used for poultry feed.

spores are spread by the wind, and grasshoppers are infected by contact.

A bait formulation of a microsporidian, *Nosema locustae* Canning, is produced commercially. It is relatively slow acting and should be applied after nymphs have emerged but before adults have developed. Surviving infected adults consume less vegetation and produce fewer eggs.

Grasshopper eggs, nymphs, and adults are attacked by many kinds of predators and parasites, including ground beetles, blister beetles, flies, wasps, threadworms, and birds.

Chemical Control

Chemicals are registered for grasshopper control and can be applied with ground or air equipment. Materials are formulated as baits or emulsifiable concentrates. Chemicals provide the rapid results necessary to manage large outbreaks.

Cultural Control

Cultural practices include clean cultivation to eliminate weedy hosts that attract ovipositing females, late season tillage to expose egg pods, and elimination of foliage required by newly hatched nymphs. Early seeding of spring crops may increase plant tolerance to defoliation.

Selected References

Capinera, J. L., and Hibbard, B. E. 1987. Bait formulations of chemical and microbial insecticides for suppression of crop-feeding grasshoppers. J. Agric. Entomol. 4:337-344.

Harris, J. L. 1985. Grasshopper Control. Saskatchewan Agriculture, Regina, Saskatchewan, Canada.

Hewitt, G. B., and Onsager, J. A. 1983. Control of grasshoppers on rangeland in the United States—A perspective. J. Range Manage. 36:202-207.

Lavigne, R. J., and Pfadt, R. E. 1966. Parasites and predators of Wyoming rangeland grasshoppers. Sci. Monogr. 3. University of Wyoming, Laramie.

Lockwood, J. A. 1993. Environmental issues involved in biological control of rangeland grasshoppers (Orthoptera: Acrididae) with exotic agents. Environ. Entomol. 22:503-518.

Parker, J. R., and Connin, R. V. 1964. Grasshoppers: Their habits and damage. U.S. Dep. Agric. Agric. Bull. 287.

Putnam, L. C. 1962. The damage potential of some grasshoppers of the native grassland of British Columbia. Can. J. Plant Sci. 42:596-601.

Schlebecker, J. T. 1953. Grasshoppers in American agricultural history. Agric. Hist. 27:85-93.

Chapter 10

Mormon Cricket
(Orthoptera: Tettigoniidae)

Some Important Species

Mormon crickets are large, wingless insects first encountered by early settlers in the Salt Lake area of Utah in 1848. Since then, extensive crop damage has occurred in the northern Great Plains.

Important species include *Anabrus simplex* Haldeman (Mormon cricket), *Peranabrus scabricollis* (Thomas) (coulee cricket), *A. cerciata* Caudell, and *A. longipes* Caudell.

General Characteristics

Description

Adult Mormon crickets are large and wingless. Colors range from light gray to dark brown. They have gradual life cycles and chewing mouthparts. The female has a prominent, swordlike ovipositor.

Distribution

Mormon crickets occur only in North America and have been found from Manitoba to Arizona and west to California.

Life Cycle

Female crickets (Plate 14) insert individual eggs into the soil during the summer. Eggs overwinter, and nymphs (Plate 15) emerge during the following spring, undergo seven instars, and are mature in about 60 days.

Mass dispersal of crickets is common, and individuals may move as far as 2 km per day.

Fig. 10.1. Bait spreader constructed from an automobile axle.

Fig. 10.2. Flame thrower drawn by horses and used to kill Mormon crickets.

Cricket populations are suppressed during cold, wet springs because of the slow growth of the nymphs and the increased activity of insect pathogens. Outbreaks occurred during the drought of the 1930s, and then populations declined when annual precipitation increased.

Control

Biological Control

Adults and nymphs are attacked by birds, skunks, lizards, and mice. Nematodes and ground beetles feed on eggs. Naturally occurring biological agents are important pest control factors when popu-

Fig. 10.3. Flame thrower and boom used for Mormon cricket control.

lation densities of crickets are low but are generally ineffective when large outbreaks occur. An early invasion was brought under control by flocks of hungry sea gulls, and a gull statue that commemorates the event still stands in Salt Lake City, Utah.

Chemical Control

Baits consisting of arsenic or sodium fluorosilicate mixed with wheat bran were applied during early outbreaks and were sometimes applied with homemade spreaders (Fig. 10.1). Crickets were also killed with horse-drawn flame throwers that were fueled with oil (Figs. 10.2–10.4).

Fig. 10.4. Herding crickets into range of the flame thrower.

Fig. 10.5. Metal fence used to exclude Mormon crickets from a wheat field.

Fig. 10.6. Mound of crickets killed in an oil-covered moat.

Cultural Control

Some of the control practices used by early settlers took advantage of the crickets' inability to fly. Trenches were dug around fields to impede invading crickets. Some trenches were filled with water and coal oil. Also, short metal fences were erected around fields (Figs. 10.5 and 10.6). Other practices included plowing to kill eggs and burning to kill nymphs and adults. Herds of sheep were driven over infested fields to crush the crickets.

Related Species

Field crickets, *Acheta* spp. (Plate 16), are common in fields late in the growing season. They are scavengers and seldom cause crop damage.

Selected References

Cowan, F. T. 1990. The Mormon cricket story. Mont. State Univ. Agric. Exp. Stn. Spec. Rep. 31.

Wakeland, C. 1959. Mormon crickets in North America. U.S. Dep. Agric. Tech. Bull. 1202.

Chapter 11

Thrips
(Thysanoptera: Thripidae)

Some Important Species

Thrips are small, cigar-shaped insects commonly found in small grain. Feeding by nymphs and adults on leaves causes pale streaks.

Important species include barley thrips, *Limothrips denticornis* Haliday; grain thrips, *L. cerealium* (Haliday); and flower thrips, *Frankliniella tritici* (Fitch).

General Characteristics

Description

Adult thrips are dark brown or black, and some have white markings. The margins of the wings are fringed with hairs. Adult males may be wingless. Nymphs (Plate 17) are white or pale green. Thrips are small and reclusive and jump or fly when disturbed. They undergo modified complete metamorphosis and have piercing and chewing mouthparts used to rasp plant surfaces. Some species are important predators of mites and small insects. Magnification and knowledge of the group are required for identification of thrips.

Distribution

Thrips occur across agricultural regions of North America, Europe, and Asia.

Life Cycle

Females overwinter in sod or other sheltered areas and disperse to crops during the spring. Eggs are laid in host plants, and hatching occurs in 10–12 days. Thrips are unusual in that the nymphs resemble adults, although there is an inactive pupal stage. Individuals of some species undergo two nymphal stages, one prepupal stage, and a pupal stage before reaching maturity in 30–35 days. There are one to five generations per year.

Detection, Sampling, and Monitoring

Thrips may be collected with sweep nets in the field and from plant debris samples processed in Berlese funnels. Plants may be shaken or tapped over sheets of white paper to dislodge and detect thrips. Nymphs and adults are commonly found under leaf sheaths. Plants may also be collected and placed in plastic bags for dissection and examination in the laboratory.

Estimations of population densities are complicated by the small size and rapid movement of thrips. Characteristic leaf streaking and associated fecal spots are indicators of infestation.

Crop Damage and Losses

Thrips feed on small grains, fruits, and flowers. Young plants are attacked in the spring and fall. Feeding occurs within the sheath when plant heads are forming. Pollen, flowers, and developing grain are attacked. Infestations reduce the quantity and quality of grain.

Economic Thresholds

Two adults and 30 immature insects per plant caused a loss of 2.5 bushels per acre in North Dakota.

Control

Insecticides are registered for use and are effective for control.

Selected References

Bates, B. A., and Weiss, M. J. 1991. The spatial distribution of *Limothrips denticornis* Haliday (Thysanoptera: Thripidae) eggs on spring barley. Can. Entomol. 123:205-210.

Bates, B. A., Weiss, M. J., Carlson, R. B., and McBride, D. K. 1991. Sequential sampling plan for *Limothrips denticornis* (Thysanoptera: Thripidae) on spring barley. J. Econ. Entomol. 84:1630-1634.

Post, R. L., and McBride, K. D. 1966. Barley thrips biology and control. Univ. N. D. Circ. A-292.

Chapter 12

Chinch Bugs
(Heteroptera: Lygaeidae)

Some Important Species

Chinch bugs attack small grains, corn, and grasses. Important species include the chinch bug, *Blissus leucopterus leucopterus* (Say); the hairy chinch bug, *B. leucopterus hirtus* Montandon; and the western chinch bug, *B. occiduus* Barber.

General Characteristics

Description

Newly laid eggs are white but become reddish. Young nymphs are yellow, but they turn red as they grow. Adult bugs are black with white forewings. Wings of adult chinch bugs fold back over the abdomen. Adults and nymphs have piercing and sucking mouthparts.

Distribution

Chinch bugs are found from Mexico to Canada.

Life Cycle

Adult bugs overwinter in wild grasses, woodlots, and other protected areas. Eggs are laid in host plants during the spring. Annual reproduction begins in small grains, and the insects disperse to corn when the small grain crop matures. There are two to three generations per year. They undergo gradual metamorphosis.

Detection, Sampling, and Monitoring

Nymphs and adults may be found by inspecting stems of host plants.

Crop Damage and Losses

Nymphs and adults suck juices from small grain and corn plants. Infested plants may be stunted or killed.

Chinch bug survival is favored by hot, dry weather. Heavy rains cause high mortality rates in young nymphs. Overwintering adults may be killed by low temperatures. This pest has been economically important in the past but currently is considered to be of minor importance.

Control

High humidity during the summer encourages attack by the fungus *Beauveria bassiana* (Balsamo). A parasitic wasp, *Eumicrosoma beneficum* Gahan, attacks chinch bug eggs. Predators include big-eyed bugs, birds, and lady beetles.

Control practices include use of resistant crops such as alfalfa and soybeans in rotation programs. Insecticides are available to control outbreaks. Barriers made from soil or paper and lined with creosote have been used to prevent nymphs from migrating to corn.

Selected References

Gossard, H. A. 1911. The chinch bug. Ohio Agric. Exp. Stn. Circ. 115.

Packard, C. M. 1951. How to fight the chinch bug. U.S. Dep. Agric. Farmers' Bull. 1780.

Schaber, B. D. 1992. Chinch bug reappears after 20 years. Agric. Can. Weekly Newsl. 3028.

1. Wheat stem broken by hail resulting in bleached stem and head.

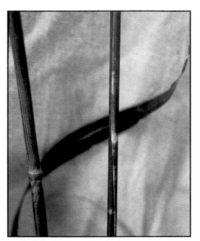

2. Pock mark on wheat stem caused by hail.

3. Wheat stems bent by hail.

4. Wheat head shattered by hail.

5. Adult grasshopper attracted from dry rangeland to green wheat plants.

6. Cone-headed grasshopper.

7. Spur-throated grasshoppers make up most of the economically important groups.

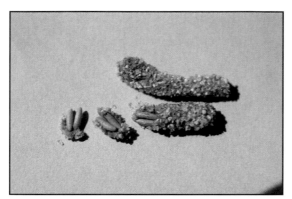

8. Grasshopper eggs and egg pods.

9. Newly emerged grasshopper nymph.

10. Wheat defoliation caused by grass-hoppers.

11. Wheat head damage caused by grasshoppers.

12. Wheat stem damage caused by grasshopper feeding.

13. Grasshopper feeding damage on the edge of a wheat field.

14. Mormon cricket adult female. Colors range from brown and black to gray and green.

15. Mormon cricket nymph feeding on wild grass heads.

16. Adult field cricket scavenging on plant material.

17. Thrips nymphs are yellow. Adults are black.

18. Adult Say stink bug feeding on developing wheat kernels.

19. Bird cherry-oat aphid feeds by sucking juice from plants.

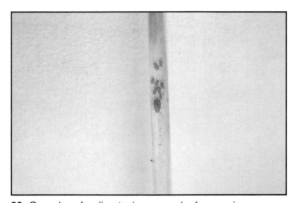

20. Greenbug feeding toxin causes leaf necrosis.

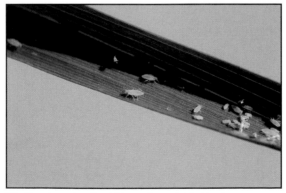

21. Russian wheat aphids prefer unfolded leaf areas.

22. Colonies of winged female aphids and young nymphs on a wheat leaf.

23. White streaks on a wheat leaf caused by Russian wheat aphid feeding.

24. Goosenecked wheat head resulting from honeydew accumulations on the flag leaf sheath.

25. Beneficial damsel bug feeding on an aphid.

26. Predatory lady beetles attracted by aphids.

27. Lady beetle larva feeding on an aphid.

28. Parasitoid attacking aphids.

29. Parasitized aphid "mummies."

30. Wheat head armyworm moth resting on a wheat leaf.

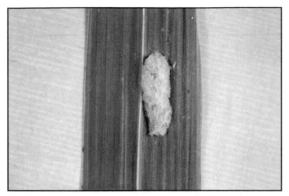

31. Fall armyworm egg masses covered with scales.

32. Fall armyworm larvae emerging from eggs.

33. Nearly full-grown fall armyworm larva.

34. Cutworm pupa in earthen cell.

35. Dead cutworm covered with fungal spores.

36. Cutworm killed by a virus.

37. Young wheat seedling damaged by army cutworm.

38. Pale western cutworms feeding underground.

39. Wheat head armyworm attacking developing kernels.

40. Wheat kernels damaged by the wheat head armyworm.

41. Army cutworm with hundreds of parasite cocoons.

42. Cereal leaf beetle and eggs on a wheat leaf.

43. Cereal leaf beetle larvae feeding on wheat.

44. Damage to corn caused by cereal leaf beetle feeding.

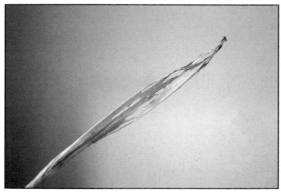

45. Damage to wheat caused by cereal leaf beetle feeding.

46. Click beetle (adult wireworm).

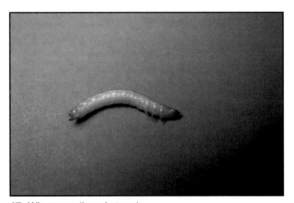

47. Wireworm (larval stage).

48. Excavation to detect wheat seedlings killed or damaged by wireworms.

49. Uneven wheat growth caused by wireworm feeding.

50. Dead and damaged wheat stems resulting from wireworm feeding.

51. Holes in seedling leaves caused by wireworm feeding.

52. Plant infected by *Cephalosporium gramineum* at a wireworm feeding site.

53. Left, false wireworm and right, true wireworm.

54. False wireworm larva.

55. False wireworm adult.

56. Barley seedlings cut below the soil surface by false wireworms.

57. Wheat stem sawfly adult.

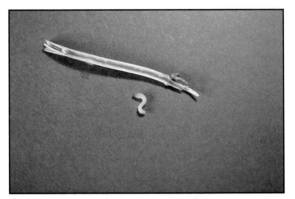

58. Wheat stem sawfly larva extracted from a wheat stem.

59. Overwintering site of sawfly larvae in underground regions of wheat stems.

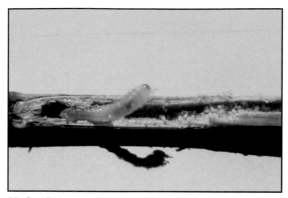

60. Sawfly larva and frass, or "sawdust," in a wheat stem.

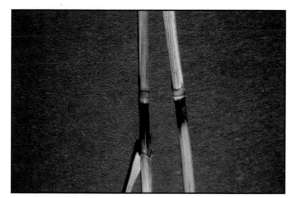

61. Wheat stem discoloration caused by accumulation of carbohydrates below nodes damaged by feeding sawfly larvae.

62. Wheat stem lodging caused by sawfly girdling.

63. Wasp emergence hole in the top of a wheat stub.

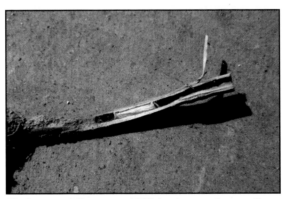

64. Cocoon of *Bracon cephi* (Gahan), a sawfly parasite.

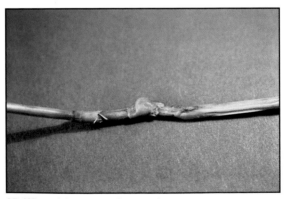

65. Wheat jointworm gall on a wheat stem.

66. Newly emerged Hessian fly female releasing a sex pheromone.

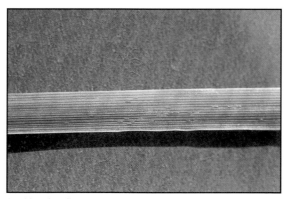

67. Hessian fly eggs on a wheat leaf.

68. Hessian fly larvae on a wheat seedling stem.

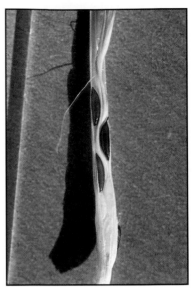

69. Hessian fly larvae and puparia under the sheath of a wheat plant.

70. Left, Hessian fly puparium and right, seed of flax

71. Hessian fly pupal skins protruding from a wheat plant after flies have emerged.

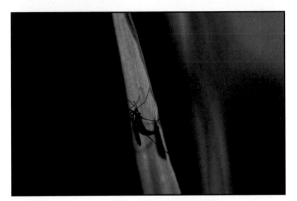

72. Hessian fly laying eggs on a wheat leaf.

73. Hessian fly puparia and a pupal skin.

74. Wheat seedling killed by Hessian flies.

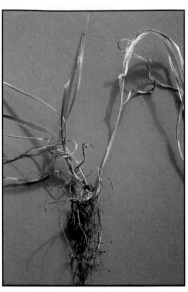

75. Wheat tiller killed by Hessian flies.

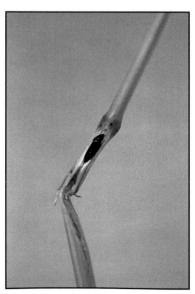

76. Weakened wheat stem infested by Hessian flies.

77. Hessian fly larvae and puparia concealed under wheat sheath.

78. Bottom, wheat head of stem infested with Hessian flies; top, wheat head of uninfested stem.

79. Wheat stem maggot adult.

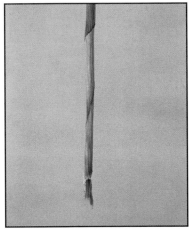

80. Stem cutting caused by wheat stem maggot.

81. End of stem chewed by wheat stem maggot.

82. White head caused by the wheat stem maggot.

83. Grass sheathminer fly and puparia in leaf sheaths.

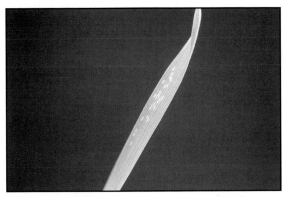

84. Wheat leaf with oviposition scars caused by the grass sheathminer.

85. June beetle eggs and a larva.

86. Plant damage and silk-lined underground tunnel of the lesser cornstalk borer.

87. "Dead heart" caused by the frit fly.

88. Tiger moth.

89. Tiger moth caterpillar.

90. Leaf sawfly larva.

91. Billbug grub.

92. Plant bug.

93. Plant bug.

94. Plant bug feeding damage.

Chapter 13

Say Stink Bug
(Hemiptera: Pentatomidae)

Stink bugs, *Chlorochroa sayi* Stål, are occasionally pests in small grains.

Several other species of phytophagous stink bugs occur in crops. At least one species is predaceous on cutworms and other pest insects. "Sunn pests" are economically important stink bug pests in Europe and could become important in North America if they were accidentally introduced.

General Characteristics

Description

Adults are large, triangular insects. Green summer coloration (Plate 18) fades to brown or gray in the fall. Stink bugs are moderately strong fliers, and wings fold backward over the abdomen. They undergo gradual metamorphosis. Small, cylindrical eggs are laid in groups on plant surfaces. Nymphs and adults have piercing and sucking mouthparts.

Life Cycle

Adults overwinter in plant debris or under weeds in the field. Eggs are laid on host plants, and nymphs are full grown in 3–4 weeks. There is one generation per year in the north and two or more in the south.

Detection, Sampling, and Monitoring

Bugs are easily detected on plants and can be collected with sweep nets.

Crop Damage and Losses

Nymphs and adults feed on weeds early in the growing season and disperse to cereal grain fields when kernels begin to fill. Loss of plant solutions and injection of toxic salivary enzymes cause lower yields and poor grain quality. Stink bug feeding on stems can produce sterile, sun-bleached heads.

Economic Threshold

Treatment with insecticides should be considered when more than three or four adult bugs are captured in 100 sweeps.

Control

Weeds and wild grasses can be sprayed or mowed to reduce the availability of hosts that support stink bug reproduction.

Selected Reference

Jacobson, L. A. 1965. Damage to wheat by Say stink bug, *Chlorochroa sayi*. Can. J. Plant Sci. 45:413-417.

Chapter 14

Aphids
(Homoptera: Aphididae)

Aphids are small, soft-bodied insects that reproduce rapidly. They damage crops by vectoring pathogens, injecting toxic saliva, and removing plant juices.

Some Important Species

Bird Cherry-Oat Aphid

The length of the antennae of the bird cherry-oat aphid, *Rhopalosiphum padi* (L.) (Plate 19), exceeds half the body length. There is a reddish brown band or spot near the rear of the abdomen. This aphid vectors plant viruses and produces toxic saliva.

Corn Leaf Aphid

The length of the antennae is less than half that of the body of the corn leaf aphid, *R. maidis* (Fitch). The cornicles, a pair of protruding dorsal appendages, are dark colored and readily visible. Corn is infested after small grains mature.

English Grain Aphid

The length of the antennae exceeds half the body length of the English grain aphid, *Sitobion avenae* (Fabricius). The cornicles are black. This aphid species probably is the most common in small grains in the temperate zone. Colonies develop on leaves and then move to heads after the boot stage.

Greenbug

The length of the antennae of the greenbug, *Schizaphis graminum* (Rondani) (Plate 20), is more than half the body length. There is a dark green dorsal stripe on the abdomen. The name is misleading, since this insect is an aphid and not a "bug." Toxic saliva is produced, and infested plants may be killed. There are three to four generations per month. The aphid overwinters in the southern states.

There are several biotypes. Biotype A was the original form, and plant breeders developed resistant wheat, barley, and oats. In 1965, biotype B appeared and was able to infest wheat resistant to biotype A. Biotype C appeared as a pest of resistant sorghum in 1968 and is more tolerant of high temperatures and lighter colored than biotypes A and B. It is possibly a new introduction from the Mediterranean area. Biotype D is resistant to some insecticides and morphologically resembles biotype C.

Russian Wheat Aphid

Antennae and cornicles of the Russian wheat aphid, *Diuraphis noxia* (Kurdjumov) (Plate 21), are short. There is a waxy body covering. The body is tubular, and two anterior protuberances may be seen during careful examination. This species appeared in Mexico in 1980, and the range had expanded into Canada by 1988. It overwinters as nymphs or adults throughout most of its range. Infested plants have white or purple streaks, and heavy infestations may lead to total crop loss. Losses in 1989 were estimated to be $276 million. Resistant wheat, barley, and rye cultivars are being developed. Pesticides are successfully used for control.

General Characteristics

Description

Aphids have cornicles and may be winged or wingless. They have piercing and sucking mouthparts and undergo gradual metamorphosis. Field populations frequently include more than one species.

Distribution

Aphids occur throughout the wheat-producing areas of the United States and Canada. They are prevalent in temperate regions of the world.

Life Cycle

Aphids can make unique adaptations that permit exploitation of short-lived herbaceous host plants. Alternate periods of sexual and asexual reproduction allow for genetic recombination, but the reproductive rate is maximized by the fact that only female offspring are produced when succulent summer host plants are available. Dispersal to available crops during the growing season permits selection of suitable host growth stages in asynchronous agricultural systems. Aphid infestations can appear suddenly and increase explosively.

Aphid life cycles are sometimes varied and complex. In a generalized model, winged females overwinter on woody hosts and emerge from eggs during the spring. These aphids fly to spring hosts, including small grains, where they give birth to wingless female nymphs. Temperature, crowding, decline of the host plant, and other factors encourage the birth of nymphs that develop into winged adults. Winged aphids may disperse to crops that are succulent later in the growing season. In the fall, winged males and females develop. They mate, and eggs are laid to overwinter on hosts.

There are many variations of this general pattern. For example, Russian wheat aphids are not known to produce male individuals or eggs in North America, and there are no alternate overwinter hosts.

The overwintering ability varies among aphid species. For example, the English grain aphid overwinters in western Oregon but not in the northern Great Plains. Russian wheat aphids survive in the northern border of the wheat-producing region of North America.

Summer survival during hot, dry periods also presents a problem for aphids. For example, Russian wheat aphids do not survive through the summer in southeastern Texas. Spring and fall populations result from dispersal of individuals from infestations located in the central region of the state.

Aphid dispersal is aided by wind currents. Large-scale air convection cells carry individuals upward to be returned to the ground at a new location as a result of atmospheric circulation. There are two distinct annual flight periods of the greenbug in Kansas. Annual spring dispersal also occurs from Texas and Oklahoma northward and from Washington and Oregon eastward into other small grain-producing areas where overwintering does not occur.

Detection, Sampling, and Monitoring

Colonies of aphids (Plate 22) can be found by examining leaves and stems of host plants. Young leaves should be unrolled to reveal aphids concealed within the plants. Heads and roots are also colonized. Sequential sampling models indicate that control decisions can be made after the presence or absence of aphids has been recorded for 25–76 tillers, depending on aphid population density. Fields should be scouted monthly during the fall and at weekly intervals during the spring.

Aphid movement can be monitored with sticky traps, suction traps, yellow pan traps, or aerial nets. Sweep nets are effective for detection of established populations.

Crop Damage and Losses

Aphid outbreaks have destroyed more than 50 million bushels of grain during 1 year in the United States. Small grains, sorghum, forage grasses, corn, and other crops are attacked. Individuals of various aphid species may show host preference; smooth brome, *Bromus inermis* Leysser, and big bluestem, *Andropogon gerardii* Vitman, are the least preferred.

Aphids cause stunting, leaf distortion, and leaf streaking (Plate 23). Leaves may curl or fail to unroll. Toxic saliva can retard root growth. Emerging plant heads may be bent or "goosenecked" during emergence from the boot (Plate 24). Heavy infestations can kill plants.

Aphids probe and feed in phloem tissue. Feeding also occurs on developing kernels. Damaged kernels are light and shriveled and have a lower protein content. Kernels that have reached the mid-dough stage are tolerant to feeding damage.

Feeding by the Russian wheat aphid disrupts osmoregulatory processes, resulting in loss of turgor and droughtlike plant symptoms. Fall feeding reduces the ability of the host to survive the winter. The Russian wheat aphid infests about 63% of the wheat and 58% of the barley in the United States.

Aphids release large amounts of honeydew, which adheres to plant surfaces. Honeydew supports fungal growth that may interfere with sunlight reception and reduce photosynthetic activity. Sticky plants also may cause problems during harvest.

The feeding mechanism and dispersal ability of aphids make them ideal candidates for transmitting plant pathogens such as the causal agents of maize dwarf mosaic and barley yellow dwarf. Aphids that have fed on infected plants produce less honeydew and more winged offspring.

Economic Thresholds

Economic thresholds are affected by many factors. Some aphid species are more toxic to host plants than others. Wheat is more sensitive than rye to feeding damage, and early plant growth stages are less tolerant than later ones.

The economic injury level for English grain aphids in spring wheat is two to four aphids per tiller at flowering, six to 10 aphids at the milky ripe stage, and more than 10 aphids from the milky to the medium dough stage. The economic injury level of Russian wheat aphids is two to four aphids per seven plants in the fall or one to two aphids per seven plants in the spring. Winter wheat should be scouted at monthly intervals during the fall and every 2 weeks when growth resumes during the spring.

Control

Biological Control

Aphids are vulnerable to many kinds of predators and parasites. Predators include damsel bugs (Plate 25) and lady beetles (Plates 26 and 27). Early season development of alfalfa and associated insects support development of predators that later move to small grains and then to corn.

Six species of hymenopterous parasites are known to attack aphids on small grains in Idaho. A native parasite, *Lysiphlebus testaceipes* (Cresson), attacks the greenbug. Parasites are very small and difficult to detect (Plate 28). The occurrence of paperlike aphid "mummies" (Plate 29) indicates parasite activity.

Although parasites reproduce rapidly and may bring aphid populations under control, predatory insects commonly are outnumbered because of the fecundity of aphids. Therefore, predators may suppress aphids when population levels are low, but heavy aphid infestations may not be suppressed quickly enough to prevent crop losses.

Chemical Control

Pesticides are registered for use and are effective for aphid control. Treatment decisions should be based on population dynamics of aphids and beneficial insects.

Cultural Control

Volunteer grain and grassy weeds should be controlled to reduce hosts, especially between harvest and the appearance of fall-seeded crops. Late fall planting will reduce chances of infestations.

Host Plant Resistance

Resistant lines of barley, oats, wheat, and sorghum are available. Resistance is in the form of antibiosis and tolerance. Resistance in wheat has broken down in the case of greenbug biotype B. This demonstrates the ongoing need for research to develop new resistant crops.

Selected References

Ajayi, O., and Dewar, A. M. 1982. The effect of barley yellow dwarf virus on honeydew production by the cereal aphids, *Sitobion avenae* and *Metopolophium dirhodum*. Ann. Appl. Biol. 100:203-212.

Archer, T. L., and Bynum, E. D., Jr. 1992. Economic injury level for the Russian wheat aphid (Homoptera: Aphididae) on dryland winter wheat. J. Econ. Entomol. 85:987-992.

Burd, J. D., Burton, R. L., and Webster, J. A. 1993. Evaluation of Russian wheat aphid (Homoptera: Aphididae) damage on resistant and susceptible hosts with comparisons of damage ratings to quantitative plant measurements. J. Econ. Entomol. 86:974-980.

DeBarro, P. J. 1992. The role of temperature, photoperiod, crowding and plant quality on the production of alate viviparous females of the bird cherry-oat aphid, *Rhopalosiphum padi*. Entomol. Exp. Appl. 65:205-214.

Feng, M. G., Johnson, F. B., and Talbert, S. E. 1992. Parasitoids (Hymenoptera: Aphidiidae and Aphelinidae) and their effect on aphid (Homoptera: Aphididae) populations in irrigated grain in southern Idaho. Environ. Entomol. 21:1433-1440.

George, K. S., and Gair, R. 1979. Crop loss assessment on winter wheat attacked by the grain aphid, *Sitobion avenae* (F.). Plant Pathol. 28:143-149.

Gildow, F. E. 1980. Increased production of alatae by aphids reared on oats infected with barley yellow dwarf virus. Ann. Entomol. Soc. Am. 73:343-347.

Girma, M., Wilde, G. E., and Harvey, T. L. 1993. Russian wheat aphid (Homoptera: Aphididae) affects yield and quality of wheat. J. Econ. Entomol. 86:594-601.

Girma, M., Wilde, G. E., and Reese, J. C. 1992. Russian wheat aphid (Homoptera: Aphididae) feeding behavior on host and nonhost plants. J. Econ. Entomol. 85:395-401.

Greene, G. L. 1966. Biology studies of *Macrosiphum avenae* (Fabr.), *Acyrthosiphon dirhodum* Walker, and *Rhopalosiphum padi* (L.) on graminae in western Oregon. Ph.D. diss. Oregon State University, Corvallis.

Johnson, R. L., and Bishop, G. W. 1987. Economic injury levels and economic thresholds for cereal aphids (Homoptera: Aphididae) on spring-planted wheat. J. Econ. Entomol. 98:478-482.

Kieckhefer, R. W., and Kantack, B. H. 1980. Losses in yield in spring wheat in South Dakota caused by cereal aphids. J. Econ. Entomol. 73:582-585.

Kieckhefer, R. W., and Kantack, B. H. 1988. Yield losses in winter grains caused by cereal aphids (Homoptera: Aphididae) in South Dakota. J. Econ. Entomol. 81: 317-321.

Mayo, Z. B., Jr., and Starks, K. J. 1974. Temperature influences an alary polymorphism in *Schizaphis graminum*. Ann. Entomol. Soc. Am. 67:421-423.

Montandon, R., Slosser, J. E., and Frank, W. A. 1993. Factors reducing the pest status of the Russian wheat aphid (Homoptera: Aphididae) on wheat in the rolling plains of Texas. J. Econ. Entomol. 86:899-905.

Niassy, A., Ryan, J. D., and Peters, D. C. 1987. Variations in feeding behavior, fecundity, and damage of biotypes B and E of *Schizaphis graminum* (Homoptera: Aphididae) on three wheat genotypes. Environ. Entomol. 16:1163-1168.

Olsen, C. E., Pike, K. S., Boydson, L., and Allison, D. 1993. Keys for identification of apterous viviparae and immatures of six small grain aphids (Homoptera: Aphididae). J. Econ. Entomol. 86:137-148.

Peters, D., Kerns, D., Puterka, G. J., and McNew, R. 1988. Feeding behavior, development, and damage by biotypes B, C, and E of *Schizaphis graminum* (Homoptera: Aphididae) on 'Wintermalt' and 'Post' barley. Environ. Entomol. 17: 503-507.

Rose, A. H., Silversides, R. H., and Lindquist, O. H. 1975. Migration flight by an aphid, *Rhopalosiphum maidis* (Hemiptera: Aphididae), and a noctuid, *Spodoptera frugiperda* (Lepidoptera: Noctuidae). Can. Entomol. 107:567-576.

Schuster, D. J., and Starks, K. J. 1975. Preference of *Lysiphlebus testaceipes* for greenbug resistant and susceptible small grain species. Environ. Entomol. 4: 887-888.

Starks, K. J., and Burton, R. L. 1977. Greenbugs: A comparison of mobility on resistant and susceptible varieties of four small grains. Environ. Entomol. 6:331-332.

Starks, K. J., and Burton, R. L. 1977. Greenbugs: Determining biotypes, culturing, and screening for plant resistance with notes on rearing parasitoids. U.S. Dep. Agric. Tech. Bull. 1556.

Starks, K. J., Burton, R. L., and Merkle, O. G. 1983. Greenbugs (Homoptera: Aphididae) plant resistance in small grains and sorghum to biotype E. J. Econ. Entomol. 76:877-880.

Webster, J. A., Porter, D. R., Baker, C. A., and Mornhinweg, D. W. 1993. Resistance to Russian wheat aphid (Homoptera: Aphididae) in barley: Effects on aphid feeding. J. Econ. Entomol. 86:1603-1608.

Chapter 15

Armyworms and Cutworms (Lepidoptera: Noctuidae)

Armyworms and cutworms are the larval stages of night-flying moths or millers. Many kinds of crops are attacked, and moths can be nuisances in homes when they are attracted to lights.

Some Important Species

There are several hundred species of cutworms and armyworms in North America, but damage to small grains is caused by six species. Cutworms are difficult to identify. If possible, representative specimens should be reared to obtain moths for identification. There is a considerable amount of variation in moth and larval coloration within species, and the appearance of male and female moths may vary greatly within species.

Army Cutworm

The eggs of the army cutworm, *Euxoa auxiliaris* (Grote), hatch in the fall, and partially grown larvae overwinter. Cutworms feed on tips of leaves during the spring and may complete their development on winter grains before spring grains appear. Moths (Fig. 15.1A) emerge in early summer, migrate to high elevations, and return to prairies to oviposit during the fall. Grizzly bears feed on aggregations of army cutworm moths that accumulate in alpine habitats. When population densities are high, larvae may deplete food sources and disperse for several miles. There is one generation per year in northern regions. The lower portion of the reniform spot on the forewing of the moth is darker than the upper portion. Larvae have dark, horizontal skin pigmentation.

Pale Western Cutworm

The eggs of the pale western cutworm, *Agrotis orthogonia* Morrison (Fig. 15.1A), are laid in the fall and hatch in the spring. Cutworms remain underground and cut plants below the soil surface. The larval skin is shiny and clear and lacks prominent hairs. There is one generation per year in northern regions.

Redbacked Cutworm

The eggs of the redbacked cutworm, *E. ochrogaster* (Guenée), overwinter, and infestations sometimes accompany those of the pale western cutworm. Eggs are laid in soil in weedy stubble. Larvae are

Fig. 15.1. Guide to identification of some moths of economic importance in small grains. A, Top, pale western cutworm; bottom, Army cutworm (note that the lower halves of the reniform spots are darker than the upper halves). B, Redbacked cutworm. Note the dark pigmentation on the forewings. C, Armyworm. Note the small white spot in the center of each forewing. D, Fall armyworm (male). Note the light coloration at the wing tips.

light pink, and moths (Fig. 15.1B) have dark red pigmentation on the forewings. There is one generation per year in northern regions.

Wheat Head Armyworm

Larvae from the first generation of the wheat head armyworm, *Faronta diffusa* (Walker), feed on foliage, and second generation armyworms feed on developing kernels. Kernel damage is similar in appearance to that caused by weevil feeding in grain bins. Larvae feed during the day and can be collected with sweep nets. They have white, green, and brown longitudinal stripes. There are two generations per year in northern regions.

Armyworm

Pseudaletia unipuncta (Haworth) is sometimes called the "true" armyworm. There are several generations per year. Larvae may disperse in "armies" when food sources are depleted. Eggs are laid in foliage. Armyworms feed at night and rest on lower leaves or on the ground during the day. Moths (Fig. 15.1C) have a prominent, tiny white spot in the center of each forewing.

Fall Armyworm

Spodoptera frugiperda (J. E. Smith), the fall armyworm (Fig. 15.1D), overwinters in southern Texas and Florida and disperses northward after producing multiple generations during the growing season. The name was assigned as a result of its late-season appearance in New York. Stems, foliage, and heads of many plants are attacked.

General Characteristics

Description

Moths are gray, and the wings are covered with scales (Plate 30). Wingspans range from 2 to 5 cm. Eggs are tiny and round and are laid either individually or in masses in the soil or on host plants (Plate 31). Young cutworms (Plate 32) may feed gregariously or disperse and develop individually. The names "armyworm" and "cutworm" are synonymous.

Cutworms have chewing mouthparts and hardened head capsules. Bodies are soft, flexible, and sparsely covered with hairs (Plate 33). Legs are found on three anterior body segments, four midbody segments, and one anterior segment. The coloration of cutworms varies within species and is influenced by crowding, age, and the food source. Full-grown cutworms are 2–4 cm long. The brown, shiny pupae are usually found in cells in the soil (Plate 34).

Distribution

Cutworms occur throughout North America. Climatic zones, vegetation types, and geographic topography affect the distribution of species and seasonal biology.

Life Cycle

Northern species of cutworms mate in late summer and begin to lay eggs in the soil during late August or early September. One moth may lay several thousand eggs. Oviposition continues until the first killing frost. Egg hatching may occur during the fall or be delayed until the following spring.

Cutworms feed for 6–8 weeks and are 2–4 cm long when full grown. Cutworm feeding and development are faster at high temperatures. Mature cutworms enter the soil and form pupal cells, which resemble clods of dirt and are difficult to find.

Some species produce about one generation per month during the growing season. Moths of univoltine species emerge during early summer, estivate through hot, dry conditions, and resume activity during the fall.

Climate affects survival and development of cutworms. Moisture is necessary for survival of eggs. However, wet conditions enhance pathogen activity (Plates 35 and 36). Also, cutworms are forced to the surface in water-saturated soil, where they are exposed to parasites and predators.

Outbreaks are promoted by 1) moderately dry conditions in midsummer, which favor moth survival; 2) moderately moist conditions during the fall, which favor egg hatch, cutworm survival, and growth of wild flowers that are nectar sources for moths; 3) delayed fall freeze, which permits completion of egg laying; and 4) low numbers of predators and parasitoids.

Detection, Sampling, and Monitoring

Young cutworms chew small holes in leaves. This feeding symptom can be detected by careful plant inspection. Most feeding and defoliation is caused by large larvae, and as a result, damage may not be apparent until larval development and feeding is nearly completed. Some cutworms feed on weeds as well as on small grains.

Most species of cutworms can be found during the day by searching through the dry upper layer of soil near damaged plants. A small hand rake is useful for locating larvae within plant rows. Other collection methods include pitfall traps and solar bait stations. They can also be found under burlap bags placed on the ground.

The extent of moth activity during the fall is an indication of the potential for cutworm outbreaks during the following spring. Activity can be monitored with light traps or sex pheromone traps. Male moths are attracted to the pheromones, but the traps cannot be used for control purposes. Pheromones are available commercially for some economically important species. Traps can be made from polyvinyl chloride pipes. Sticky-surfaced traps can become saturated during heavy flights and are less satisfactory.

Crop Damage and Losses

Cutworms feed on small grains, corn, sorghum, forage grasses, and weeds. Larvae occurring at low population densities may select succulent, preferred plant parts, but at high densities, all green vegetation is consumed.

Most species of economically important cutworms are foliage feeders. Plant seedlings may be devoured early in the growing season (Plate 37). Aboveground feeders may consume leaves and then burrow downward to consume stems below the soil surface. Plants may be severely stunted or killed. Later in the growing season, leaves of tillered or jointed plants may be partially consumed, and plants may outgrow the damage. Some species of cutworms cut plants at the surface of the soil (Plate 38), and although some material is consumed, much remains uneaten on the soil surface. These plants may continue to grow or may send up new tillers.

Cutworms may also feed on soft, developing kernels (Plate 39). Damaged, hollow kernels are light in weight and may be lost during harvest (Plate 40).

Differences in the biology of individual cutworm species affect the extent of crop damage. Eggs of most species are laid during the fall. Some hatch before winter and overwinter as larvae. As a result, overwintering larvae feed early in the spring on young crops. In contrast, larvae that overwinter as eggs appear later when crops are more mature.

Spring growth of infested fields is uneven because of the uneven distribution of the feeding cutworms. Damage may appear first on south-facing slopes and is similar to damage caused by wireworms.

Economic Thresholds

Treatments should be considered when pale western cutworms are present at rates of one per 240 cm of row in young plants or one per 120 cm in tillered plants. Army cutworm treatment is considered if there are more than two per 120 cm of row in young plants or four per 120 cm in tillered plants.

Fields with infestation levels below the action thresholds should not be treated to prevent higher infestation levels from appearing during the following year. Moths may disperse from infested fields, and mortality of full-grown cutworms may preclude development of moths. After all eggs have hatched, cutworm population densities cannot increase until moths lay more eggs and produce another generation.

Crop losses are dependent on many factors. Plant tolerance of defoliation is reduced by drought stress. Young plants are more likely to be killed than older plants. The various cutworm species cause different amounts of crop loss resulting from preferential feeding on leaves, stems, and plant heads. The relative age of cutworms and plant age are important in the extent of damage that occurs. The partially grown cutworms feed on newly emerging wheat during the spring and cause more damage than young cutworms that feed on large, well-developed plants near the end of the growing season.

Control

Biological Control

Cutworms are attacked by many kinds of predators, including ground beetles, ants, birds, and mammals.

Dipteran and hymenopteran parasites of many species attack cutworms. Unfortunately, parasites usually do not kill larvae until crop damage has occurred. In fact, *Copidosoma bakeri* (Howard) is a commonly occurring polyembryonic hymenopteran parasite that causes host larvae to feed longer and grow larger than unparasitized cutworms (Plate 41). The increase in the prevalence of parasites and predators resulting from cutworm outbreaks is an important factor in suppressing future outbreaks.

Bacterial and viral pathogens are more prevalent during wet periods. Commercial strains of *Bacillus thuringiensis* subsp. *kurstaki* are available and registered for use in small grains. The toxins are specific for lepidopterans, and there are no known environmental hazards associated with the use of this material.

Chemical Control

Cutworm damage occurs quickly, and heavily infested fields must be treated to prevent total crop loss. Newly developed insecticides are effective at very low doses and have minimal effect on the environment and nontarget organisms when used according to the labels. Fortunately, these materials are effective even during the periods of low temperature common during spring infestations.

Cultural Control

Fields that are destroyed by cutworms early in the growing season may be replanted if reseeding is delayed until cutworms have starved or pupated.

Selected References

Bauernfeind, R. J., and Wilde, G. E. 1993. Control of army cutworm (Lepidoptera: Noctuidae) affects wheat yields. J. Econ. Entomol. 86:159-163.

Burton, R. L., Starks, K. J., and Peters, D. C. 1980. The army cutworm. Okla. State Univ. Agric. Exp. Stn. Bull. B-749.

Byers, J. R. 1992. Difference in weight gain during final stadium of pale western and army cutworms related to life history and crop damage. Can. Entomol. 124:515-520.

Byers, J. R., and Struble, D. L. 1987. Monitoring population levels of eight species of noctuids with sex-attractant traps in southern Alberta, 1978–83: Specificity of attractants and effect of target species abundance. Can. Entomol. 119:541-556.

Byers, J. R., and Yu, D. S. 1993. Parasitism of the army cutworm, *Euxoa auxiliaris* (Grt.) (Lepidoptera: Noctuidae), by *Copidosoma bakeri* (Howard) (Hymenoptera: Encyrtidae) and effects on crop damage. Can. Entomol. 125:329-335.

Capinera, J. L., and Schaefer, R. A. 1983. Field identification of adult cutworms, armyworms, and similar crop pests collected from light traps in Colorado. Colo. State Univ. Coop. Ext. Serv. Bull. 514A.

Cooley, R. A., and Parker, R. J. 1916. Pages 97-107 in: The Army Cutworm in Montana. Mont. Agric. Exp. Stn. Circ. 52.

Gerber, G. H., Walkof, J., and Juskiw, D. 1992. Portable, solar-powered charging system for blacklight traps. Can. Entomol. 124:553-554.

Morrill, W. L. 1988. Evaluation of new pheromone trap designs for cutworm moths (Lepidoptera: Noctuidae). J. Econ. Entomol. 81:735-737.

Steck, W., and Bailey, B. K. 1978. Pheromone traps for moths: Evaluation of cone trap designs and design parameters. Environ. Entomol. 7:449-455.

Story, R. N., and Keaster, A. J. 1982. Development and evaluation of a larval sampling technique for the black cutworm (Lepidoptera: Noctuidae). J. Econ. Entomol. 75:604-610.

Turnock, W. J. 1987. Predicting larval abundance of the Bertha armyworm, *Mamestra configurata* Wlk., in Manitoba from catches of male moths in sex pheromone traps. Can. Entomol. 119:167-178.

Chapter 16

Cereal Leaf Beetle
(Coleoptera: Chrysomelidae)

Oulema melanopus (L.), cereal leaf beetles and grubs, feed on the foliage of small grains, corn, and grasses. Heavily damaged fields have a "frosted" appearance.

General Characteristics

Description

Beetles are about 50 mm long. Elytra are shiny blue green, and the thorax is orange (Plate 42). Antenna segments are uniform in width. The larvae, or grubs, have black heads and yellow bodies (Plate 43) and are covered with dark, slimy fecal material. The shiny orange eggs are visible on plant foliage.

Distribution

This pest is a native of Europe and was detected in Michigan in 1959. Infestations had extended northward into Quebec by 1965 and eastward into West Virginia by 1968. Dispersal to the west did not extend farther than Minnesota because of unfavorable environmental conditions.

Isolated populations appeared in Utah in 1984 and in Montana in 1988. Long-range dispersal of beetles may occur in straw. Fields are commonly infested during harvest, and beetles are found in harvested grain. However, they are unable to survive for more than 14 days in storage bins.

Life Cycle

Adults become active during the spring when temperatures exceed 16 C and feed on wild grasses near overwinter sites. Beetles fly to fields during April, mate, and feed for about 2 weeks. Each female beetle lays several hundred eggs on host plant leaves over an 8-week period. Eggs hatch in 5 days, and grubs feed on leaves and mature in about 10 days. Mature grubs leave the plants and build pupal cases in the top 5 cm of soil. After 2–3 weeks, new beetles emerge from the soil and feed for about 3 weeks on late-maturing small grains, corn, or weeds. Feeding damage in corn does not penetrate the entire leaf (Plate 44). Beetles leave the fields after harvest and are in obligatory diapause. Adults overwinter in wooded areas or other protected locations. There is one generation per year.

Factors that influence population densities include changes in overwinter habitat, climate-related mortality, high soil temperatures during pupation, and changes in cropping systems.

Detection, Sampling, and Monitoring

The shiny, orange eggs can be found by careful examination of the surfaces of host plant leaves. Grubs are also found on leaf surfaces and can be collected with sweep nets. The characteristic between-vein feeding by grubs is easily seen. Pupae are found in the soil and probably could be collected with flotation techniques. Beetles can be collected with sweep nets. Beetles also feed between leaf veins, but unlike grubs, they feed completely through the leaves of small grains. Defoliation of this type indicates that beetles are or were present. Heavy feeding damage results in dull, gray foliage (Plate 45).

Crop Damage and Losses

Beetles and grubs prefer to feed on leaves of oats and wild oats. Winter and spring wheat, barley, corn, and other grasses are also attacked, but alfalfa is resistant.

Economic Thresholds

Grubs and beetles skeletonize leaf surfaces and reduce the photosynthetic area of host plants. In addition, water loss by host plants is increased, and sites are created for pathogen infection. Damaged plants senesce earlier, produce fewer tillers, and have reduced yields.

Losses can be estimated by marking infested and uninfested plants during the growing season and then comparing plant yields. Beetles or grubs can be caged on plants to establish the effects of various infestation rates at selected plant growth stages.

Leaf areas damaged by grubs can be measured with a leaf-area meter. A grub consumes 20% of the surface of a single leaf of an oat plant, and the associated loss of grain is influenced by the stage of development of the plant. Young plants are the most sensitive to defoliation. Wheat losses range from 2 to 4 kg/ha per larva per stem. Significant losses occur in wheat when flag leaf damage exceeds 50%. Infestation levels in barley averaging 1.6 grubs per stem reduce yields by 38%.

Economic injury levels are three or more eggs or grubs per stem before the boot stage or one or more grubs per flag leaf during heading in wheat. One larva per stem in wheat or 1.5 larvae in barley or oats will reduce yields by a value of $22 per hectare.

Control

Biological Control

Five species of exotic parasites have been released and are established in North America. These include an egg parasite, *Anaphes flavipes* (Foerster), which produces many generations per year, and a larval parasite, *Tetrastichus julis* (Walker), which produces two generations per year. Although these parasites are well established, their importance is not clear. Beetle populations began declining before parasite populations reached levels high enough to impact infestations. Parasites from populations in the eastern states are established in Utah and have been recently released in Montana.

Mass rearing of parasites in laboratories has been unsuccessful. Field insectaries consisting of one planting of winter wheat and three plantings of spring oats in a location protected from spring flooding and insecticide drift can be established to enhance parasite survival. When larvae reach population densities of 43 per square meter, parasites are released. Parasites overwinter near the surface of the soil; therefore, plots should not be tilled. Parasitized larvae should be available for distribution 3–5 years after establishment of the insectary.

Chemical Control

Several chemicals are effective for controlling eggs, grubs, and beetles. The effect of insecticides on parasites can be minimized by applying treatments between parasite generations or before spring activity begins.

Host Plant Resistance

Wheat, oats, and barley lines have been screened for resistance to larval feeding, and resistance in the form of leaf pubescence has been identified. Leaf pubescence causes fewer eggs to be laid and deters feeding by young grubs. Resistant cultivars are not commercially available.

Selected References

Baniecki, J. F., and Weaver, J. E. 1972. The cereal leaf beetle in West Virginia. W. Va. Univ. Agric. Exp. Stn. Curr. Rep. 60.

Burger, T. L. Undated. Cereal Leaf Beetle Parasitoid Field Manual. U.S. Department of Agriculture, Animal and Plant Health Inspection Service, Niles, MI.

Casagrande, R. A., Ruesink, W. G., and Haynes, D. L. 1977. The behavior and survival of adult cereal leaf beetles. Ann. Entomol. Soc. Am. 70:19-30.

Connin, R. V., and Hoopingarner, R. A. 1971. Sexual behavior and diapause of the cereal leaf beetle, *Oulema melanopus*. Ann. Entomol. Soc. Am. 64:655-660.

Doward, K. 1963. European cereal pest found in Michigan. Agric. Chem. 17:59.

Haynes, D. L., and Gage, S. H. 1981. The cereal leaf beetle in North America. Annu. Rev. Entomol. 26:259-287.

Jensen, G. 1970. Cereal leaf beetle update. Page 5 in: Mont. Crop Health Rep. 5.

Karren, J. B. 1989. Cereal leaf beetle in Utah. Utah State Univ. Mimeogr. Rep.

Morrill, W. L., Jensen, G. L., Weaver, D. K., Gabor, F. W., and Lanier, W. T. 1992. Cereal leaf beetle (Coleoptera: Chrysomelidae): Incidence at harvest and survival in storage in Montana. J. Entomol. Sci. 27:1-4.

Puttler, D. I. Undated. North Carolina cereal leaf beetle parasitoid field insectaries. N. C. Dep. Agric. Mimeogr.

Webster, J. A. 1983. Cereal leaf beetle [*Oulema melanopus* (L.)] population densities and winter wheat yields. Crop. Prot. 2:431-436.

Webster, J. A., and Smith, D. H., Jr. 1979. Yield losses and host selection of cereal leaf beetles in resistant and susceptible spring barley. Crop Sci. 19:901-904.

Wellso, S. G., Reusink, W. G., and Gage, S. H. 1975. Cereal leaf beetle: Relationships between feeding, oviposition, mating, and age. Ann. Entomol. Soc. Am. 68:663-668.

Wilson, M. C., Treece, R. E., Shade, R. E., Day, K. M., and Stivers, R. K. 1969. Impact of cereal leaf beetle larvae on yields of oats. J. Econ. Entomol. 62:699-702.

Chapter 17

Wireworms
(Coleoptera: Elateridae)

Larvae are called wireworms, and adults are called click beetles. Wireworms feed on underground roots and stems of many kinds of plants. Infested crops show irregular growth and lack vigor. Damage is frequently misdiagnosed as winter kill.

Some Important Species

Important species include the Great Basin wireworm, *Ctenicera pruinina* (Horn); the western field wireworm, *Limonius infuscatus* Motschulsky; the dryland wireworm, *C. glauca* (Germar); and the prairie grain wireworm, *C. aeripennis destructor* (Brown).

General Characteristics

Description

Adults are flat, cigar-shaped beetles. They are brown, black, or bronze, and sizes vary with species (Plate 46). Beetles' heads can be snapped or "clicked" to startle predators or turn upright.

Wireworm eggs are tiny, round, and white. Newly hatched larvae are soft and white but become hard, shiny, and yellow when they grow. The head capsule is generally the same color as the body. There are three pairs of small legs. The smooth, wiry, cylindrical body contributes to ease of movement through the soil (Plate 47).

Distribution

Wireworms occur throughout agricultural regions of North America. The distribution of species is affected by rainfall, soil type, and vegetation.

Life Cycle

Beetles develop from pupae during late summer but do not emerge from the soil until the following spring. Males appear first and quickly mate with emerging females. Dispersion flights are known for some species.

Females burrow underground and lay eggs. They deposit about 350–400 eggs during a 3-week period. Eggs hatch in 3–4 weeks, and newly emerged larvae feed on plant rootlets or decaying plant material. They seldom cause damage in the first year. The larval stage may last as long as 12 years. The pupal stage lasts for about 3 weeks during late summer, and pupae are found in earthen cells 6–12 cm below the soil surface.

Detection, Sampling, and Monitoring

Wireworms can be found in infested fields by searching through the soil near damaged plants (Plate 48). Feeding damage may be found by excavating and examining dead plants. Wireworms are sometimes difficult to find because they move deep underground when temperatures rise and soil moisture decreases.

Wireworm population densities may be estimated by evaluating soil samples collected from the plant root zone. Soil can be shaken or washed through wire screens to retrieve wireworms. Wireworms are attracted to the warm, moist soil conditions created by solar bait stations. The stations can be constructed during the fall and examined during the spring to determine whether wireworms are present and seed should be treated.

Eggs can be collected by washing soil through a series of pans with varying sizes of fine mesh wire screens. Pitfall and emergence traps are useful to determine seasonal activity of beetles.

Crop Damage and Losses

Wireworms attack many crops including small grains, potatoes, onions, lettuce, beans, sugar beets, and corn. Some species prefer the fibrous roots of grasses.

Damage to crops is caused by wireworms that feed underground on seed, roots, and stems. Damaged seeds may fail to germinate, and seedlings may be killed. Infested fields have uneven plant growth

(Plate 49), and early spring damage may appear on southern slopes first.

Some tillers may be killed (Plate 50), and leaves of surviving tillers may have ragged holes resulting from wireworm feeding (Plate 51). Damage to winter wheat is enhanced when plant tolerance is reduced by drought stress or when there is no snow cover to protect plants from low temperatures. Infested fields may be destroyed or yields significantly reduced.

Wireworms feed during the spring when soil temperatures reach 7 C. There is relatively more feeding during cool, wet springs. Wireworms stop feeding and move deep underground when soil temperatures rise and moisture levels drop. They also feed during the fall.

Larval feeding damage on stems may increase the incidence of the fungus *Cephalosporium gramineum,* which appears as brown discoloration (Plate 52).

Grasslands are important reservoirs of wireworms. An estimated 200,000 beetles per hectare can be produced annually. This figure is based on emergence trap catches in Georgia. Crop damage is more likely to occur on land that previously was in sod.

Control

Biological Control

There are no applied biological control practices for wireworms. Naturally occurring predators include birds, toads, and insects. Wireworms are cannibalistic in captivity.

Chemical Control

Wireworms and other subterranean insect pests are difficult to control. Prophylactic seed treatments should be applied in regions where wireworm damage has occurred. Yields increased by 30% in regions of a field in which treated seed was planted. The field had an infestation level of three larvae per solar bait station.

During dry periods, growers must plant seed more deeply than usual to reach moist soil. The insecticidal activity of the material used for seed treatment is restricted to the area close to the seed. Therefore, young plants may extend above the protected area in the soil and may be damaged by wireworm feeding.

Efficacy of experimental materials can be estimated by counting dead and damaged plants per unit area. Soil must be dug and examined because there may be no aboveground evidence of dead plants. Stand counts should be made before tillering. Weakened and dead wireworms may accumulate in treated rows.

Cultural Control

Cultural control practices include clean summer fallow, fall tillage to crush pupae, and flooding to kill larvae. Seeding rates can be increased to compensate for seedling loss. Healthy, well-fertilized plants may outgrow damage.

Related Species

Damage caused by wireworms is similar to that caused by false wireworms. False wireworms generally are larger and have a shorter life cycle and therefore cause more damage over a shorter period of time. Spring damage caused by cutworms and armyworms may occur at the same time as that caused by wireworms.

Selected References

Doane, J. F. 1977. Spatial pattern and density of *Ctenicera destructor* and *Hypolithus bicolor* (Coleoptera: Elateridae) in soil in spring wheat. Can. Entomol. 109:807-822.

Fisher, J. R., Keaster, A. J., and Fairchild, M. L. 1975. Seasonal vertical movement of wireworm larvae in Missouri: Influence of soil temperature on the genera *Melanotus* Escholtz and *Conoderus* Escholtz. Ann. Entomol. Soc. Am. 68: 1071-1073.

Lane, M. C. 1931. The Great Basin wireworm in the Pacific Northwest. U.S. Dep. Agric. Farmers' Bull. 1657.

Morrill, W. L. 1978. Emergence of click beetles (Coleoptera: Elateridae) from some Georgia grasslands. Environ. Entomol. 7:895-896.

Morrill, W. L. 1984. Wireworms: Control, sampling methodology, and effect on wheat yield in Montana. J. Ga. Entomol. Soc. 19:67-71.

Onsager, J. A. 1969. Sampling to detect economic infestations of *Limonius* spp. J. Econ. Entomol. 62:183-189.

Shirck, F. H. 1930. A soil-washing device for use in wireworm investigations. J. Econ. Entomol. 23:991-994.

Shirck, F. H. 1942. The flight of sugar-beet wireworm adults in southwestern Idaho. J. Econ. Entomol. 35:423-427.

Strickland, E. H. 1939. Life cycle and food requirements of the northern grain wireworm, *Ludius aereipennis destructor* Brown. J. Econ. Entomol. 32:322-329.

Toba, H. H., O'Keeffe, L. E., Pike, K. S., Perkins, E. A., and Miller, J. C. 1985. Lindane seed treatment for control of wireworms (Coleoptera: Elateridae) on wheat in the Pacific Northwest. Crop Prot. 4:372-380.

Toba, H. H., and Turner, J. E. 1983. Evaluation of baiting techniques for sampling wireworms (Coleoptera: Elateridae) infesting wheat in Washington. J. Econ. Entomol. 76:850-855.

Treherne, R. C. 1923. Wireworm control. Can. Dep. Agric. Pamph. 33.

Ward, R. H., and Keaster, A. J. 1977. Wireworm baiting: Use of solar energy to enhance early detection of *Melanotus depressus, M. verberans*, and *Aeolus mellillus* in Midwest cornfields. J. Econ. Entomol. 70:403-406.

Chapter 18

False Wireworms
(Coleoptera: Tenebrionidae)

False wireworms feed on underground stems and roots of plants. Damage and biology are similar to those of true wireworms.

Some Important Species

Important species include *Eleodes extricata* (Say), *E. hispilabris* (Say), *E. opacus* (Say) (the plains false wireworm), *E. suturalis* (Say), *E. tricostata* (Say), and *Embaphion muricatum* Say.

General Characteristics

Description

False wireworms resemble true wireworms but have longer legs and more conspicuous antennae and move more rapidly (Plates 53 and 54). They inhabit the low-rainfall areas of western North America.

Adult false wireworms are large, black beetles that move about on the soil surface (Plate 55). Individuals of some species raise their posteriors and release noxious fluids when they are threatened by predators. Some are nocturnal or crepuscular and hide in cracks or small holes in the soil during the day. They cannot fly.

Seven species are associated with crops in South Dakota. There are differences among species in preferred slopes and soil types and in seasonal development.

Distribution

False wireworms are more prevalent in arid wheat-producing regions.

Life Cycle

Most species overwinter as either larvae or adults. Eggs may be laid throughout the growing season, although *Embaphion muricatum* eggs are found in the spring and *E. extricata* eggs are laid in the fall.

Detection, Sampling, and Monitoring

Adult false wireworms are active on the soil surface and can readily be captured in pitfall traps. Larvae can be found by searching through the soil and sometimes appear on the soil surface after heavy rains.

Crop Damage and Losses

False wireworms were present in native prairie grasses and were of economic importance early in the history of agriculture in the Great Plains. Larvae feed on seeds and underground portions of young plants of wheat, oats, barley, rye, and corn. Damage appears as skips in rows or circular areas in fields that lack vigorous plant growth. Damage is commonly misdiagnosed as winter kill and can be confused with damage caused by cutworms. Beetles also feed on seeds and plants but have not caused significant crop losses.

The greatest amount of crop damage occurs in the fall, especially when wheat germination is delayed by an unusually low amount of rainfall. Seeds are consumed, and underground portions of stems of young plants may be destroyed (Plate 56).

Damage is commonly associated with sandy or sandy loam soils. The extent of crop damage is enhanced by the presence of straw stacks or other material that attracts ovipositing beetles. Weedy or trashy fields and consecutive wheat production in the same field for 2 years or more promotes infestations. False wireworm population densities are currently increasing because of a reduction in acreages of alternate-year summer fallow.

Economic Thresholds

Accurate estimations of crop damage are difficult because plants may compensate for death of neighboring plants. However, population densities of three to four larvae per 35 cm of plant row damaged 60% of wheat seeds, and significant yield loss occurred when an average of one larva per 30 cm of plant row was present when plants were in early growth stages.

Control

Biological Control

Parasitoids include *Perilitus eleodis* Viereck; a sarcophagid, *Sarcophaga eleodis* Aldrich; and tachinids. Predators including mice and skunks are repelled by defensive secretions released through the anal glands. Other predators include ground beetles (*Calasoma* spp., *Harpalus* spp., and *Pasimachus* spp.), crows, meadowlarks, and blackbirds.

Chemical Control

Seed should be treated in areas where there is a history of crop damage. Fungicides and insecticides are commonly combined as seed treatments.

Cultural Control

Clean summer fallow and crop rotation interrupt life cycles and keep populations below economically important levels.

Selected References

Allsopp, P. G. 1980. The biology of false wireworms and their adults (soil-inhabiting Tenebrionidae) (Coleoptera): A review. Bull. Entomol. Res. 7:343-379.

Calkins, C. O., and Kirk, V. M. 1973. Distribution and movement of adult false wireworms in a wheat field. Ann. Entomol. Soc. Am. 66:527-532.

Calkins, C. O., and Kirk, V. M. 1975. False wireworms of economic importance in South Dakota (Coleoptera: Tenebrionidae). U.S. Dep. Agric. Agric. Res. Serv. B633.

Morrill, W. L., Lester, D. G., and Wrona, A. E. 1990. Factors affecting efficacy of pitfall traps for beetles (Coleoptera: Carabidae and Tenebrionidae). J. Entomol. Sci. 25:284-293.

Chapter 19

Wheat Stem Sawfly
(Hymenoptera: Cephidae)

The wheat stem sawfly, *Cephus cinctus* Norton, has adapted from native grasses to wheat and become the most destructive, consistent insect pest in the northern Great Plains. Infestations cause reduction in grain production and extensive losses resulting from lodging during harvest.

General Characteristics

Description

Wasps are black and about 1 cm long. They have bright yellow markings on the abdomen (Plate 57). The wings are clear but appear golden in the sunlight. Eggs are very small but are visible without magnification when plant stems are dissected. Larvae have brown head capsules and cream-colored bodies and assume an S shape when removed from stems (Plate 58). Tunnels in stem stubs are lined with a thin, silken sheath that encloses overwintering larvae and pupae.

Distribution

Populations of economic importance occur in Montana, North Dakota, Manitoba, Alberta, and Saskatchewan. Wheat stem sawflies have been collected throughout western North America.

Life Cycle

Sawflies overwinter as mature larvae in the stubs or lower portions of stems (Plate 59). Tunnels in the stubs extend downward several centimeters below the soil surface to the plant crowns. The underground environment provides protection from extreme temperatures, desiccation, and parasitoids.

Sawfly larvae are freeze resistant. In the laboratory, larvae freeze and die at −22 C. There is no seasonal preparation for overwintering; the supercooling ability of larvae is the same during the summer, fall, and spring. The silken lining of tunnels in stubs prevents contact of larvae with water or ice in stem walls. Removal of this lining reduces larval supercooling.

Pupation occurs during the spring. Wasp emergence during late May or early June coincides with stem elongation of wild grasses. Timing of wasp emergence is critical to survival because host stems must be present and must be succulent long enough to support development of sawfly larvae.

Wasps emerge during early summer by pushing through the soft plugs that fill the tops of the stubs. Wasps are able to dig through crusted soil. Cool, wet weather extends the emergence period.

Wasps fly from infested stubble to the first suitable host and begin to lay eggs. Therefore, the heaviest infestations occur in field borders. Dispersal of marked individuals was less than 2 km.

Each female wasp lays about 30 eggs. Large-diameter stems are preferred for egg laying and produce the largest larvae. Male individuals are more likely to be produced in small stems. Fertilized eggs produce females, and unfertilized eggs produce males. Stems less than 1.6 mm in diameter midway between the upper two nodes are avoided. Wheat in the boot stage is preferred. Tillering wheat is not attacked, and plants that are beginning to dry down or ripen are immune.

Sawlike ovipositors (hence the name "sawfly") are used to insert eggs into stems. Although wasps deposit only one egg before moving to another stem, more eggs may be laid in the stem by subsequent females. Eggs hatch in about 8 days. Larvae are cannibalistic, and only one will survive in hollow-stemmed hosts. Larval boring is restricted in solid-stemmed plants, and occasionally two larvae per stem will survive. There is one generation per year.

Detection, Sampling, and Monitoring

Wasps are easily seen on plant foliage, and sweep net sampling is effective for detecting wasps where there are low population densities.

Larval infestation rates can be estimated by splitting stems to detect larvae or frass (Plate 60). Darkened areas below stem nodes (Plate 61) indicate the presence of larvae, although stems should be dissected to verify infestation.

Stubble samples can be collected after harvest to determine the prevalence of infested stubs. Tops of stubs are plugged and have a characteristic clean cut (Plate 62). Stubs are often below the soil surface; therefore, excavation is necessary to obtain all that are present.

Wasp emergence dates can be estimated by monitoring development of pupae during the spring. Emergence holes in stub plugs indicate that wasps have left (Plate 63).

Plant lodging is the most obvious symptom of wheat stem sawfly infestation. However, the extent of plant lodging is not a precise indication of larval infestation rates. Many infested stems remain standing through harvest. Lodging increases when harvest is delayed and is encouraged by stress from wind or rain.

Grain loss caused by tunneling larvae can be estimated by comparing head weights of grain from infested and uninfested stems. However, comparisons must be made among plants of similar size to compensate for higher infestation rates among large stems.

Harvest losses resulting from lodging can be estimated by collecting cut stems and associated heads remaining on the ground after harvest. Although stems are also cut by hail, cutworms, and rodents, sawfly damage is easily identified by the cleanly cut straw ends and the presence of frass in stems.

Crop Damage and Losses

Wheat is the preferred host, although barley is sometimes damaged. Large-stemmed grass hosts commonly found in road ditches, rangeland, and ravines are also infested.

Larvae feed and freely move throughout the length of hollow stems. Feeding and boring in the nodes damage phloem tissue, and nutrient transport is restricted when kernels are filling. Nutrient loss reduces grain quality and quantity. Head weights are reduced by up to 30%.

Plant lodging is the most spectacular type of damage caused by wheat stem sawflies. Larvae girdle the interior circumference of stem walls near the base of maturing host plants. Weakened stems usually lodge, and grain may be lost during harvest. Lodging is enhanced by wind, rain, or any factor that physically stresses plants. Broken stems may be supported by neighboring plants in thick, vigorous stands. Lodging is more pronounced when stands are thin and drought stressed.

Management Strategies

Sawfly populations increase about 10-fold annually. For example, infestations of 7–9% may increase to 70–80% the following year. Therefore, management practices should affect at least 90% of the individuals to be effective.

Biological Control

Sawflies originally occurred at comparatively low population densities in wild grasses where populations were suppressed by parasitoids. Although sawflies quickly adapted to wheat, parasitoids have been slower to follow. In 1955, the rate of parasitism was nearly 100% in wild grasses but seldom was above 2% in wheat. Parasites affect sawfly populations in wheat in one out of every 5 years.

Nine species of native parasites attack *C. cinctus*. Predominant species are *Bracon lissogaster* Muesebeck and *B. cephi* (Gahan). Both species are bivoltine ectoparasites. Eggs are laid near or on sawfly larvae. First generation wasps appear during June and July when sawfly wasps are present. Second generation adult parasites are present during mid-August. At that time, many sawfly larvae are in wheat stubs and escape attack. Unusually long growing seasons, extended by early, warm springs or long, cool summers, result in the availability of green, succulent wheat stems for a long period and higher survival rates of second generation parasites. Parasitism rates drop to low levels when wheat ripens early for two consecutive years. Parasite cocoons can be found in stems (Plate 64).

Parasitoids of the European wheat stem sawfly, *C. pygmaeus* (L.), were collected in Sweden, England, and Russia and released in North America. They have not been recovered from the wheat stem sawfly.

Chemical Control

Chemicals generally have not been effective for sawfly control. Contact insecticides kill wasps that are present during application, but residues are not adequate to protect against wasps that appear later. Systemic insecticides may have some efficacy against young larvae.

Cultural Control

Spring and fall tillage of infested stubble causes some sawfly mortality. Reduced tillage and chemical fallow practices enhance survival of pest insects that overwinter in stubble.

Tillage is an effective method of reducing sawfly infestations if infested stubble can be pushed to the surface of the soil. Fall tillage can kill more than 90% of overwintering larvae in exposed stubs by causing exposure to low temperatures and desiccation. The ability of emerging wasps to dig through several inches of soil makes burying stubble ineffective.

In northern arid regions, stubble should be left standing over the winter to retain snow and reduce wind erosion of the soil. However, field borders (the areas of heaviest sawfly infestation) can selectively be tilled with minimal detrimental effect on fields.

Changes in cultural practices, including strip cropping, alternate-year summer fallow, and reduced acreages of oats, have enhanced wheat stem sawfly infestations.

Harvesting

The amount of lodging increases as harvest is delayed. Heavily infested fields, or the most heavily infested portions of fields, should be harvested first. Swathing (early windrowing) reduces lodging losses but increases harvest costs. Swathing does not reduce sawfly populations because larvae have retreated to the stem bases. Growers retrieve as much of the lodged grain as possible by cutting in one direction into the leaning stems. These practices reduce lodging losses but do not compensate for losses caused by larval feeding in the stems.

Modifications in Planting Date

Late-planted spring wheat may escape wasp attack. Late planting commonly occurs as a result of extended spring rains. Delayed planting in semiarid regions is not recommended because of inefficient use of soil moisture. Late-maturing crops are also more vulnerable to hail damage.

In contrast, winter wheat can be planted early in the fall, resulting in early plant maturation during the spring and possible avoidance of ovipositing wasps. However, early planted winter wheat is susceptible to fall infestations of Russian wheat aphids and plant pathogens. Early maturing winter wheat cultivars may also escape sawfly attack.

Host Plant Resistance

Stem solidness, first developed in the cultivar Rescue in 1946, has been a major factor in reducing losses in spring wheat. New, improved spring wheat cultivars, including Chinook, Fortuna, and Lew, have been released. Solid-stemmed winter wheat lines are currently being developed.

Stems become solid when elongating internodes are filled with pith. Solidness is reduced when stems elongate during cloudy weather. Extensive losses from sawflies may occur in cultivars that are considered to be solid.

Wasps lay eggs in solid stems, but many eggs and small larvae die. In one study, larval mortality was 28% in hollow stems and 67% in solid stems. Stem solidness does not reduce prevalence of parasitoids.

Burning

Stubble burning is not an effective control method for sawflies because adequate heat is not generated to kill larvae that are underground when host plants are dry enough to burn. Burning also is a source of air pollution and causes a loss of organic material.

Destruction of Wild Grass Hosts

Road ditches, waterways, and other grassy areas are reservoirs of sawflies. Young larvae can be killed if host plants are mowed soon after egg laying is finished.

Trap Strips

Trap strips bordering wheat fields can be used to intercept dispersing wasps. Wheat stem sawfly larvae will be killed when strips are mowed. However, the resulting vegetation that remains on the ground presents a problem during subsequent planting.

Crop Rotation

Sawflies survive only in grass hosts. Incorporation of oats, alfalfa, soybeans, sunflowers, canola, and other resistant crops into rotation systems will reduce sawfly populations. Unfortunately, short growing seasons and low amounts of precipitation in the northern Great Plains severely limit selection of alternate crops.

Related Species

C. cinctus and *C. clavatus* (Norton) are Nearctic species. *C. clavatus* is a little-known species that is found in western states. The European sawfly, *C. pygmaeus*, is Holarctic and was introduced into North America before 1887. It is established in eastern states and causes some damage in wheat. The black grain stem sawfly, *Trachelus tabidus* (Fabricius), is also Holarctic and was introduced prior to 1899. It occurs in small grains in eastern states.

Selected References

Ainslie, C. N. 1920. The western grass-stem sawfly. U.S. Dep. Agric. Bull. 841.

Davis, E. G., and Knapp, R. B. 1949. Estimated losses caused by the wheat stem sawfly in 1948. U.S. Dep. Agric. Agric. Res. Serv. Bur. Entomol. Plant Quarantine Spec. Suppl. 4.

Holmes, N. D. 1977. The effect of the wheat stem sawfly, *Cephus cinctus* (Hymenoptera: Cephidae), on the yield and quality of wheat. Can. Entomol. 109:1591-1598.

Holmes, N. D., and Peterson, L. K. 1960. The influence of the host on oviposition by the wheat stem sawfly, *Cephus cinctus* Nort. (Hymenoptera: Cephidae). Can. J. Plant Sci. 40:29-46.

Holmes, N. D., and Peterson, L. K. 1962. Resistance of spring wheats to the wheat stem sawfly, *Cephus cinctus* Nort. (Hymenoptera: Cephidae). II. Resistance to the larva. Can. Entomol. 94:348-365.

Holmes, N. D., and Peterson, L. K. 1965. Swathing wheat and survival of wheat stem sawfly. Agron. J. 45:579-581.

McNeal, F. H., Berg, M. A., and Luginbill, P., Jr. 1955. Wheat stem sawfly damage in four spring wheat varieties as influenced by date of seeding. Agron. J. 47:522-525.

Morrill, W. L., Gabor, J. W., Hockett, E. A., and Kushnak, G. D. 1992. Wheat stem sawfly (Hymenoptera: Cephidae) resistance in winter wheat. J. Econ. Entomol. 85:2008-2011.

Morrill, W. L., Gabor, J. W., and Kushnak, G. D. 1992. Wheat stem sawfly (Hymenoptera: Cephidae) damage and detection. J. Econ. Entomol. 85:2413-2417.

Munro, J. A. 1945. Wheat stem sawfly and harvest loss. N. D. Agric. Exp. Stn. Bull. 7:12-16.

Munro, J. A., Post, R. L., and Knapp, R. 1947. The wheat stem sawfly as affecting yield. N. D. Agric. Exp. Stn. Bull. 10:46-51.

O'Keeffe, L. E., Callenbach, J. A., and Lebsock, K. L. 1960. Effect of culm solidness on the survival of the wheat stem sawfly. J. Econ. Entomol. 53:244-246.

Painter, R. H. 1953. The wheat stem sawfly in Kansas. Trans. Kans. Acad. Sci. 56:432-434.

Platt, A. W., Farstad, C. W., and Callenbach, J. A. 1948. The reaction of Rescue wheat to sawfly damage. Sci. Agric. 28:154-161.

Seamans, H. L. 1945. A preliminary report on the climatology of the wheat stem sawfly (*Cephus cinctus* Nort.) on the Canadian prairies. Sci. Agric. 25:432-457.

Seamans, H. L., Manson, G. F., and Farstad, C. W. 1944. The effect of wheat stem sawfly (*Cephus cinctus* Nort.) on the heads and grain of infested stems. Pages 10-15 in: Annu. Rep. Entomol. Soc. Ont., 75th.

Weiss, M. J., and Morrill, W. L. 1992. Wheat stem sawfly (Hymenoptera: Cephidae) revisited. Am. Entomol. 38:241-245.

Weiss, M. J., Morrill, W. L., and Reitz, L. L. 1987. Influence of planting date and spring tillage on the wheat stem sawfly. Mont. AgResearch 4(1):5.

Weiss, M. J., Riveland, N. R., Reitz, L. L., and Olson, T. C. 1990. Influence of resistant and susceptible cultivar blends of hard red spring wheat on wheat stem sawfly (Hymenoptera: Cephidae) damage and wheat quality parameters. J. Econ. Entomol. 83:255-258.

Youtie, B. A., and Johnson, J. B. 1988. Association of the wheat stem sawfly with basin wild rye. J. Range Manage. 41:328-331.

Chapter 20

Wheat Jointworm (Hymenoptera: Eurytomidae)

Plants infested with the wheat jointworm, *Tetramesa tritici* (Fitch), have hardened nodes, reduced head weight, and increased incidence of lodging. Larvae are found in stem galls.

General Characteristics

Description

Adults are tiny, black wasps. They are difficult to find in the field. Characteristic plant symptoms indicate infestations are present.

Distribution

The wheat jointworm first appeared as a pest in wheat-producing areas east of the Mississippi River and was reported in northwestern states in 1926.

Life Cycle

Wasps emerge from infested stubble and lay eggs in stems of wheat. Up to 15 eggs may be laid in a single stem. Eggs hatch in 14 days, and larval feeding causes formation of gall tissue at the nodes (Plate 65). Larvae puncture and feed on cells within the galls. Pupae overwinter, and wasps appear in May. There is one generation per year.

Crop Damage, Detection, Sampling, and Monitoring

Wheat is the preferred host, but barley and rye are also attacked. There are few recent reports of extensive damage. Infested stems have hardened, thickened nodes and reduced head weight and grain quality.

Reduced-tillage practices could enhance survival of this pest.

Control

Cultural Control

Plowing infested stubble and utilization of nonhost crops in rotation programs have reduced populations of this pest.

Biological Control

Several parasitic wasps, including *Eupelmus* spp. and *Eurytoma* spp., attack jointworms.

Selected References

Chamberlin, T. R. 1928. The wheat jointworm in Oregon. Oreg. Agric. Exp. Stn. Circ. 86.

Knowlton, G. F., and Janes, M. J. 1933. Distribution and damage by jointworm flies in Utah. Utah Exp. Stn. Bull. 243.

Phillips, W. J. 1918. The wheat jointworm and its control. U.S. Dep. Agric. Farmers' Bull. 1006.

Chapter 21

Hessian Fly
(Diptera: Cecidomyiidae)

The Hessian fly, *Mayetiola destructor* (Say), is found in major wheat-producing areas of the United States and Canada. Maggots produce feeding toxins that kill or stunt wheat. Resistant cultivars are available.

General Characteristics

Description

Hessian fly adults resemble mosquitoes (Plate 66). Females can be identified by the faint orange abdominal coloration. Male flies are smaller and have more prominent antennae.

The tiny, orange eggs can be found on upper leaf surfaces (Plate 67) between leaf hairs (Fig. 21.1) and are barely visible without magnification. Newly emerged maggots are also orange but soon become white. A green dorsal streak consisting of plant material is visible after maggots begin to feed (Plate 68). Larvae do not have legs or hardened head capsules.

The outer skin of mature maggots turns brown and hardens (Plate 69) to form a puparium or flaxseed (Plate 70). Pupae are formed inside the puparia. The empty pupal skins sometimes protrude from the plant after flies emerge (Plate 71).

Distribution

The first Hessian flies appeared on Long Island near the camp of Hessian troops after the Revolutionary War. Since then, the pest has spread into all major wheat-producing areas. There may have been subsequent introductions of Hessian flies.

Life Cycle

Hessian flies may produce several generations per year. Fall generations appear in winter wheat, and spring generations attack both winter and spring wheat. The number of annual life cycles varies with geographic location. Reproduction continues throughout the growing season in Georgia, but only one annual generation occurs in Montana. There may be one or more spring generations followed by an estivation period during the hot, dry summer period. Activity resumes during the fall, and there may be more than one fall generation.

Maggots overwinter as larvae in puparia in infested stubble, volunteer grain, or fall-seeded grain. Flies emerge in the spring and lay eggs. Male flies emerge a day or two before females. Mating occurs soon after females appear.

Eggs are laid on upper leaf surfaces. Eggs that are laid on undersides of leaves drop from plants. Female flies face upward while

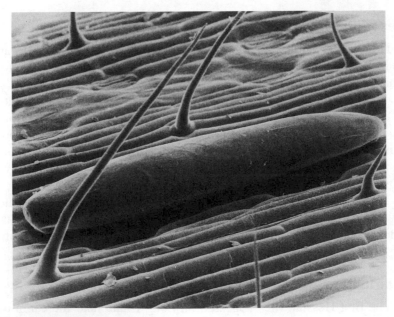

Fig. 21.1. Hessian fly egg on a wheat leaf.

they are laying eggs (Plate 72). Ovipositors have sensory organs used for selection of suitable oviposition sites (Fig. 21.2).

Newly emerged maggots move downward on the upper leaf surface toward the stems. They crawl between the leaf sheath and stem and continue downward until a node is reached. Maggots that move toward the leaf tips perish. Maggots are fragile and susceptible to desiccation.

Maggots attach to stems under leaf sheaths. After the feeding period, larvae pupate within the puparia, which remain in plants after flies have emerged (Plate 73).

Detection, Sampling, and Monitoring

Seedlings with thickened, darkened leaves should be selected for examination during field inspections. Other symptoms of infestation include dead seedlings (Plate 74), dead tillers (Plate 75), and broken or lodged stems (Plate 76). Maggots and puparia are concealed under leaf sheaths (Plate 77).

Flies are small and difficult to detect in the field, although they can be collected with sweep nets and sticky traps. Fall surveys are useful in predicting regions where infestation may occur in the next year but are inadequate for predicting the extent of damage.

Fig. 21.2. Sensory organs on a Hessian fly ovipositor.

Crop Damage and Losses

Wheat is the preferred host, but barley, rye, quackgrass, and various species of wheat grasses and rye grasses are also attacked. Oats are resistant. Infestation of wild and domestic grasses has caused foreign embargoes of exported hay.

Loss estimations can be made by comparing grain from infested stems with that from uninfested stems (Plate 78). Also, systemic insecticides can be applied at various rates to produce varying degrees of infestation. Yields can be compared at the resulting infestation levels.

Maggots feed by sucking juice from plants. Stems are not chewed, but plant tissue is stretched (Figs. 21.3 and 21.4). Plant damage is a result of feeding toxins. The growth stage at which the plant is attacked determines the resulting symptoms. Seedlings may be killed. Tillers of older plants may be killed, but new tillers may be generated. After jointing, stems are weakened and head weights are reduced.

Hessian fly outbreaks occur rapidly. In 1978, damage appeared in

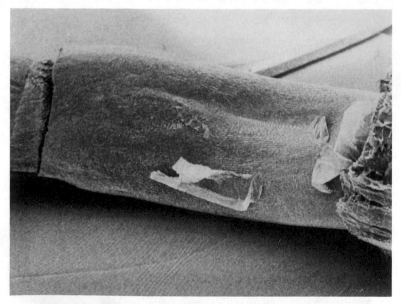

Fig. 21.3. Depression on a wheat stem where a Hessian fly maggot was located.

spring wheat in South Dakota in the largest outbreak in the United States in the past 40 years. The first brood appeared during late April to mid-May, and the second brood appeared during early July. Population densities are enhanced by reduced tillage and cool, moist springs.

Control

Biological Control

Several species of Hymenopteran parasites are important in suppressing infestations. Parasite prevalence can be determined by examining flaxseeds for presence of viable larvae or evidence of parasites. There are no recommended practices for enhancing parasite populations.

Chemical Control

Preventative systemic granular insecticides may be applied at planting in fields where consistent fly damage occurs. Efficacy of granules is reduced by low precipitation. No chemicals are effective when infestations are discovered in existing crops.

Fig. 21.4. Wheat stem tissue stretched by Hessian fly maggot feeding.

Cultural Control

Delayed fall planting may result in avoiding flies from the fall generation, but the safe late planting date in southern Georgia is unreliable. Rotation with nonhost crops such as soybeans or corn will reduce infestations. Volunteer grain and grassy weeds should be destroyed in the fall.

Host Plant Resistance

Host plant resistance is an important control practice. Resistance is in the form of larval antibiosis, meaning that young maggots cannot survive on resistant plants. Hessian flies vary in their ability to survive on wheat with different types of genes for resistance and have been separated into biotypes.

Wheat currently has 20 types of resistance genes. Examples of Hessian fly biotypes and resistant cultivars include the Great Plains biotype, found in the central United States, which cannot infest wheat with any genes for resistance. Biotype A is predominant in the eastern soft wheat-producing region of the United States and cannot infest wheats with the *H3, H5, H6,* or *Marquillo* genes for resistance. Biotype B is predominant in the east-central United States and cannot survive on wheat with the *Marquillo, H5,* or *H6* genes. Biotype E occurs in Georgia and cannot survive on wheat with the *H5* or *Marquillo* genes. Other fly biotypes have been identified in the field and the laboratory.

Growers must select wheat cultivars that are resistant to the type of Hessian fly that is present in their location. Flies in the eastern United States cannot survive harsh environmental conditions in the Great Plains, and the Great Plains fly biotype cannot survive on the soft wheats in the eastern states.

Selected References

Buntin, G. D., Ott, S. L., and Johnson, J. W. 1992. Integration of plant resistance, insecticides, and planting date for management of the Hessian fly (Diptera: Cecidomyiidae) in winter wheat. J. Econ. Entomol. 85:530-538.

Foster, J. E., and Taylor, P. L. 1974. Estimating populations of the Hessian fly. Environ. Entomol. 3:441-445.

Gahan, A. B. 1933. The serphoid and chalcidoid parasites of the Hessian fly. U.S. Dep. Agric. Misc. Pub. 174.

Gallun, R. L. 1955. Races of the Hessian fly. J. Econ. Entomol. 48:608-609.

Gallun, R. L., and Reitz, L. P. 1971. Wheat cultivars resistant to races of Hessian fly. U.S. Dep. Agric. Agric. Res. Serv. Prod. Res. Rep. 134.

Morrill, W. L. 1982. Hessian fly: Host selection and behavior during oviposition, winter biology, and parasitoids. J. Ga. Entomol. Soc. 17:156-167.

Morrill, W. L., and Nelson, L. R. 1976. Hessian fly control with carbofuran. J. Econ. Entomol. 69:123-124.

Nelson, L. R., and Morrill, W. L. 1978. Hessian fly control in wheat by suppression of fall generations with carbofuran. Agron. J. 70:139-141.

Pike, K. S., and Antonelli, A. L. 1981. Hessian fly in Washington. Wash. State Univ. Bull. XB0909.

Steiger, D. K., Walgenbach, D. D., and Cholick, F. A. 1982. Hessian fly infestation on South Dakota spring wheats, 1978–1982. Mimeogr. Rep. to the S. D. Wheat Comm.

Yoloyama, V. Y., Hatchett, J. H., and Miller, G. T. 1933. Hessian fly (Diptera: Cecidomyiidae) control by hydrogen phosphide fumigation and compression of hay for export to Japan. J. Econ. Entomol. 86:76-85.

Chapter 22

Wheat Stem Maggot
(Diptera: Chloropidae)

Plants infested by the wheat stem maggot, *Meromyza americana* Fitch, have white, or blasted, heads. Maggots destroy upper portions of stems.

Other chloropid species in North America include *M. pratorum* Meigen and *M. saltatrix* (L.). Damage is sometimes confused with that caused by other stem borers, such as frit flies.

General Characteristics

Description

Flies are about 40 mm long and have green or yellow thoracic and abdominal stripes (Plate 79). The hind tibia are greatly enlarged. Eggs are white and visible on leaves and stems. Maggots are white and legless and have one pair of black, hooklike mouthparts. Pupae are about 50 mm long and are nearly transparent.

Distribution

Wheat stem maggots occur in most wheat-producing areas of the United States and Canada. Crop damage occurs from New York to Kansas and northward into Canada.

Life Cycle

Flies appear in fields in June. About 30 eggs per female are laid on stems or leaves. Eggs hatch in about 10 days. Maggots feed and mature in about 3 weeks. Pupation lasts about 2 weeks. Newly emerged flies leave plants, mate, and produce a second generation.

Second generation eggs are laid on flag leaves, and maggots move downward to enter the plant at the junction of the leaf and stem. Maggots make circular cuts around the stem, causing death of the head (Plate 80). Dead stems and heads do not break but become bleached white by the sun. Maggots feed and complete their development above the top node.

There are two or three generations per year. Flies leave the fields prior to harvest and lay eggs in wild grasses, and the insects overwinter as maggots.

Detection, Sampling, and Monitoring

Presence and relative abundance of flies may be determined by sweep net samples. Prevalence of eggs may be determined by inspecting plants. Maggots may be found by dissecting plant stems.

Hail, plant pathogens, and other factors also cause white heads in wheat. Therefore, proper diagnosis is necessary for accurate damage estimations. White heads can easily be pulled from plants because dead stems desiccate and shrink within the leaf sheath. Maggot feeding results in ragged stem ends (Plate 81). Maggots can be found by carefully dissecting upper stem regions.

Crop Damage and Losses

Susceptible crops include wheat, rye, and barley. Other grassy hosts include quackgrass, *Elytrigia repens*; slender wheat grass, *Elymus trachycaulus,* and western wheat grass, *E. smithii*; brome grass, *Bromus inermis* and *B. japonicus*; green foxtail, *Setaria viridis,* and yellow foxtail, *S. glauca.*

Although maggot feeding is associated with white heads (Plate 82), early infestation may kill tillers and be undetected. However, significant losses of this type have not been reported.

White heads do not produce kernels and are a complete loss. Yield reductions resulting from death of young tillers are more difficult to estimate because vigorous plants may compensate for death of individual tillers or neighboring plants.

The percentage of head loss can be estimated by counting both white and green heads in a short length of row, such as 1 m or approximately one pace. Numbers of white heads seen during 50 or 100 paces can be used to calculate the percent infestation. Loss seldom exceeds 1%, although losses of up to 15% have been reported.

Crop injury is very conspicuous, and therefore damage is easily over-estimated.

Management Strategies

No pesticides are recommended for control. Although some host plant resistance has been identified, the expression is moderate and has not been used in development of commercial cultivars. Stubble tillage has been recommended for control, but it is ineffective because flies are not in the fields after harvest. Adequate fertilizer to produce healthy, vigorous plants may reduce losses.

Biological Control

Several species of parasites attack wheat stem maggots. *Bracon meromyzae* Gahan and *Coelinidea meromyzae* (Forbes) are the most common. Parasites spin cocoons in stems, and wasps chew exit holes in stems. Natural populations of parasites are important in maintaining pest populations at relatively low densities.

Cultural Control

Incorporation of nonhost crops such as corn, oats, and legumes into crop rotation systems is desirable for reducing losses. Destruction or grazing of volunteer grain in August and late planting of winter wheat may be effective.

Selected References

Allen, M. W., and Painter, R. H. 1937. Observations on the biology of the wheat stem maggot in Kansas. J. Agric. Res. 55:215-238.

Branson, T. F. 1971. Resistance of spring wheat to the wheat stem maggot. J. Econ. Entomol. 64:941-945.

Gilbertson, G. I. 1925. The wheat stem maggot. S. D. Agric. Exp. Stn. Bull. 217.

Kieckhefer, R. W. 1974. Seasonal appearance and movement of adult stem maggots in South Dakota cereal crops. J. Econ. Entomol. 67:558-561.

Kieckhefer, R. W., and Morrill, W. L. 1970. Estimates of loss of yield caused by the wheat stem maggot to South Dakota cereal crops. J. Econ. Entomol. 63:1426-1429.

Morrill, W. L., and Baldridge, D. E. 1980. Recropping and the rate of infestation by the wheat stem maggot. Proc. Mont. Acad. Sci. 39:25-27.

Morrill, W. L., and Kieckhefer, R. W. 1971. Parasitism of the wheat stem maggot in South Dakota. J. Econ. Entomol. 64:1129-1131.

Rockwood, L. P., Zimmerman, S. K., and Chamberlin, T. R. 1947. The wheat stem maggots of the genus *Meromyza* in the Pacific Northwest. U.S. Dep. Agric. Tech. Bull. 928.

Chapter 23

Grass Sheathminer
(Diptera: Agromyzidae)

Maggots of the grass sheathminer, *Cerodontha dorsalis* (Loew), feed in leaf sheaths of wheat and grasses. Mature maggots form brown puparia that can be confused with Hessian fly flaxseeds.

General Characteristics

Description

Adults are small, black flies. Maggots are white and have hook-like mouthparts. Fully developed maggots form brown puparia about 1 mm long within the tissue of leaf sheaths (Plate 83). These puparia are much smaller than Hessian fly flaxseeds.

Distribution

This insect occurs in North America and Europe.

Life Cycle

Flies insert eggs into leaf tissue and cause tiny, transparent scars (Plate 84). Maggots burrow through leaf tissue toward the stem and leave thin, transparent streaks in the leaves. Maggots feed in the sheath for 10–20 days, and flies emerge after 25 days. There are three generations per year. The insect overwinters in the pupal stage.

Crop Damage and Losses

No crop loss occurs. No control practices are necessary.

Selected Reference

Seamans, H. L. 1917. Wheat-sheath miner. J. Agric. Res. 9:17-25.

Chapter 24

Other Pest Insects of Small Grains

Some insect pest groups are little known or seldom encountered. Unusual conditions, such as wheat regrowth late in the season after hail damage, may encourage increases in population densities of some pests. Changes in cultural practices or continued adaptation from native grass species may result in new insect species achieving pest status.

Minor Pests

European Corn Borer

The European corn borer, *Ostrinia nubilalis* (Hübner) (Lepidoptera: Pyralidae), is an imported species that attacks corn but sometimes appears in wheat. Larvae bore into stems and reduce crop yield and quality. Stems are filled with frass, and mature larvae leave exit holes. Crop damage has been reported in New York and Georgia.

Orange Wheat Blossom Midge

The orange wheat blossom midge, *Sitodiplosis mosellana* (Géhin) (Diptera: Cecidomyiidae), lays eggs on developing heads of grasses and small grains, and maggots feed in kernels. Infested kernels are shriveled and bent. Outbreaks are exacerbated by damp weather. Losses have been reported in Canada and the northern United States. Delayed planting and chaff destruction after harvest may reduce losses.

June Beetles, May Beetles, and White Grubs

Infestations of June beetles, May beetles, and white grubs, *Phyllophaga* spp. (Coleoptera: Scarabaeidae), cause losses in small grains throughout North America. Beetles are attracted to lights at night and lay eggs in the soil. Grasslands are common reproduction sites. The grubs (Plate 85) feed on roots, and life cycles last 1–3 years. Clean cultivation may reduce attractiveness of fields to ovipositing beetles.

Lesser Cornstalk Borer

The lesser cornstalk borer, *Elasmopalpus lignosellus* (Zeller) (Lepidoptera: Pyralidae), attacks crops in southern regions of North America. Damage usually occurs in corn, although small grains are attacked. Larvae bore into the bases of plants, and plants may be killed. Larvae construct silk-lined tunnels at the bases of plants (Plate 86).

Frit Fly

The maggots of the frit fly, *Oscinella frit* (L.) (Diptera: Chloropidae), bore into stems. Feeding causes death of the upper part of infested plants, a condition known as "dead heart" (Plate 87).

Tiger Moths

Tiger moth (Lepidoptera: Arctiidae) caterpillars of several species disperse from rangeland to small grain fields and cause some defoliation (Plates 88 and 89).

Leafhoppers

Many species of leafhopper (Homoptera: Cicadellidae) are found in small grains. Some may fly long distances. The six-spotted leafhopper, *Macrosteles fascifrons* (Stål), vectors a mycoplasma that causes yellows or aster yellows.

Leaf Sawflies

Leaf sawflies (Hymenoptera: Tenthredinidae) feed on leaves of small grains. There are several species that appear throughout North America, but economic losses are unusual. They cause some defoliation but are usually held in check by parasitoids. Larvae are green with white stripes (Plate 90).

Billbugs

Billbugs (Coleoptera: Curculionidae) attack grass stolons and are sometimes pests of small grains (Plate 91).

Seed Maggot

The seed maggot, *Hylemya cerealis* (Gillette) (Diptera: Anthomyiidae), attacks grain in the soil.

Plant Bugs

Plant bugs (Hemiptera: Miridae) of many species are present in wild grasses (Plates 92 and 93). Some reproduce in cereal grains. Feeding causes irregular splotches on leaves (Plate 94).

Selected References

Barker, P. S. 1984. The orange wheat blossom midge, its distribution and the damage it caused to wheat in Manitoba during 1984. Proc. Entomol. Soc. Manit. 40:25-29.

Pivnick, K. A. 1993. Daily patterns of activity of females of the orange wheat blossom midge, *Sitodiplosis mosellana* (Géhin) (Diptera: Cecidomyiidae). Can. Entomol. 125:725-736.

Glossary

C—Celsius or centigrade ($°C = [°F - 32] \times 5/9$)
g—gram (1 g = 0.035 ounce)
ha—hectare (1 ha = 2.47 acres)
kg—kilogram (1 kg = 2.2 pounds)
kg/ha— kilograms/hectare (1 kg/ha = 0.89 pounds/acre)
m—meter (1 meter = 1.09 yards)
m^2—square meter (1 m^2 = 1.18 square yards)

abiotic—of or relating to nonliving factors, e.g., rainfall, temperature, and day length
activity density—measurement of insect activity based on the extent of movement and population density
antibiosis—mortality or limitation of normal insect growth and development caused by another organism

biotic—of or relating to living factors, e.g., predators and host plants
biotype—insect category below the species level, sometimes based on differential resistance to insecticides or ability to survive on resistant plants

chaff—by-product from harvest composed of small pieces from the heads, leaves, and stems of the crop
chlorosis—abnormal yellow or light color of plant tissue
chronic—long-term occurrence or exposure
combine—machine used to harvest cereal grain and other crops
cornicle—tubelike appendage on the anterior of aphids
crop rotation—incorporation of several kinds of crops in sequence in production systems
crown—underground portion of a plant where roots and stem join
cultivar—crop variety

dorsal—upper or top

133

efficacy—effectiveness

egg pod—group of grasshopper eggs

elytrum (pl. elytra)—one of a pair thickened or hardened front wings of a beetle that fold backward and usually cover the abdomen

emulsifiable concentrate—formulation of insecticide consisting of the active ingredient, an emulsifier that permits mixing with water, and other enhancing agents

estivation—period of insect inactivity during hot, dry summer periods

exoskeleton—hard outer skin of insects

exotic—introduced from other regions

feces (adj. fecal)—food and metabolic by-products that are excreted from the anus

flotation technique—method that uses water, usually with added sugar or salt to adjust specific gravity, to separate insects from soil and debris

frass—insect fecal material and chewed-up plant tissue

gall—hardened plant tissue formed in response to enzymes excreted by insects or mites

global positioning—determination of latitude and longitude by triangulation of satellites

head capsule—hardened area that covers the head of a larva

herbicide—chemical applied to kill weeds

honeydew—sticky solution excreted by aphids

hymenopteran parasite—wasp that lays eggs in, on, or near its host

insect-day—estimation of insect feeding based on the number of insects and number of days of feeding

insecticide—chemical used to kill insects

internode—region of a grass stem between hardened joints or nodes

joint—hardened region of a grass stem; also, location of leaf sheath attachment

label restrictions—information printed on pesticide label that includes application rates, crops that can be treated, and days that must elapse before harvest

larva (pl. larvae)—immature stage of insects that undergo complete metamorphosis

leaf sheath—region of a grass leaf below the blade that circles the stem

lodging—breakage of mature stems; causes difficulty during harvest because heads may be on the ground

molt—process of shedding skin by insects

mummies—brown, paperlike remains of aphids that have been killed by parasites

node—hardened joint of a grass stem

oviposit (n. oviposition)—to lay eggs
ovipositor—specialized apparatus used to lay eggs

panicle—grass plant head
parasitoid—insect parasite
pesticide—chemical applied to kill pests
pheromone—chemical released by insects to signal or communicate with other insects
phloem—vascular tissue of plants that carries solutions
plant pathogen—organism that causes plant disease
preharvest interval—duration between insecticide application and the time a crop can be harvested
prophylactic treatment—treatment applied before insects are detected to prevent crop damage
pupa (pl. pupae)—insect life stage between larva and adult
puparium (pl. puparia)—hardened outer skin of some fly larvae
PVC—polyvinyl chloride; a hard plastic commonly used for pipes

refugium (pl. refugia)—area of retreat for insects
reniform—kidney-shaped

saline seep—accumulation of salt on the soil surface resulting from the re-surfacing and evaporation of ground water
sawdust—slang for chewed plant material and insect fecal material; usually found in plant stems
senescence—decline of plants caused by maturation or stress
simulation—artificial creation of a situation, such as removal of plant foliage to mimic insect feeding
skeletonize—to remove leaf epidermis between veins, e.g., by insect feeding
small grain—oats, barley, wheat, rice, rye, or other crops grown for human and animal food
spring wheat—wheat planted during the spring season and harvested during that calendar year
stand—number of plants per unit of area
straw—stems, usually after harvest
strip cropping—planting of long, narrow fields oriented across the direction of prevailing winds
stub—short, lower region of a wheat stem in which wheat stem sawfly larvae overwinter
stubble—lower portion of plants that remain standing after harvest
summer fallow—practice of not planting cropland during 1 year to accumulate soil moisture in semiarid regions

supercooling—relating to the ability of insects to survive at temperatures below freezing

swathing—cutting small grain crops and making windrows in which plants continue to dry before harvest

systemic insecticide—insecticide that is absorbed into plants

thorax (adj. thoracic)—one of three major body regions of adult insects; located between the head and abdomen

tiller—secondary stem that compliments the primary stem of small grains

turgor—rigidity of plants resulting from internal water pressure

volunteer growth—plants that grow from seed usually lost during harvest

wheat bran—outer covering of the seed that is removed in the milling process

windowpane—transparent epidermis that remains on leaves after some types of insect feeding

winter wheat—wheat planted during the fall and harvested during the following calendar year

Index

HARPER ADAMS AGRICULTURAL
LIBRARY

COLLEGE